Statistics
Essentials

by Deborah J. Rumsey, PhD

A Wiley Brand

Statistics Essentials For Dummies®

Published by: **John Wiley & Sons, Inc.,** 111 River Street, Hoboken, NJ 07030-5774, www.wiley.com

Copyright © 2019 by John Wiley & Sons, Inc., Hoboken, New Jersey

Published simultaneously in Canada

For general information on our other products and services, please contact our Customer Care Department within the U.S. at 877-762-2974, outside the U.S. at 317-572-3993, or fax 317-572-4002. For technical support, please visit www.wiley.com/techsupport.

Wiley publishes in a variety of print and electronic formats and by print-on-demand. Some material included with standard print versions of this book may not be included in e-books or in print-on-demand. If this book refers to media such as a CD or DVD that is not included in the version you purchased, you may download this material at http://booksupport.wiley.com. For more information about Wiley products, visit www.wiley.com.

Library of Congress Control Number: 2019937507

ISBN 978-1-119-59030-9 (pbk); ISBN 978-1-119-59024-8 (ebk); ISBN 978-1-119-59023-1 (ebk)

10 9 8 7 6 5 4 3 2 1

Contents at a Glance

Table of Contents

Introduction

This book is designed to give you the essential, nitty-gritty information typically covered in a first semester statistics course. It's bottom-line information for you to use as a refresher, a resource, a quick reference, and/or a study guide. It helps you decipher and make important decisions about statistical polls, experiments, reports, and headlines with confidence, being ever aware of the ways people can mislead you with statistics, and how to handle it.

Topics I work you through include graphs and charts, descriptive statistics, the binomial, normal, and t-distributions, two-way tables, simple linear regression, confidence intervals, hypothesis tests, surveys, experiments, and of course the most frustrating yet critical of all statistical topics: sampling distributions and the Central Limit Theorem.

About This Book

This book departs from traditional statistics texts and reference/ supplement books and study guides in these ways:

>> **Clear and concise step-by-step procedures** that intuitively explain how to work through statistics problems and remember the process.

>> **Focused, intuitive explanations** empower you to know you're doing things right and whether others do it wrong.

>> **Nonlinear approach** so you can quickly zoom in on that concept or technique you need, without having to read other material first.

>> **Easy-to-follow examples** reinforce your understanding and help you immediately see how to apply the concepts in practical settings.

>> **Understandable language** helps you remember and put into practice essential statistical concepts and techniques.

Conventions Used in This Book

I refer to statistics in two different ways: as numerical results (such as means and medians) or as a field of study (for example, "Statistics is all about data.").

The second convention refers to the word *data*. I'm going to go with the plural version of the word data in this book. For example "data are collected during the experiment" — not "data is collected during the experiment."

Foolish Assumptions

I assume you've had some (not necessarily a lot of) previous experience with statistics somewhere in your past. For example, you can recognize some of the basic statistics such as the mean, median, standard deviation, and perhaps correlation; you can handle some graphs; and you can remember having seen the normal distribution. If it's been a while and you are a bit rusty, that's okay; this book is just the thing to jog your memory.

If you have very limited or no prior experience with statistics, allow me to suggest my full-version book, *Statistics For Dummies,* to build up your foundational knowledge base. But if you are someone who has not seen these ideas before and either doesn't have time for the full version, or you like to plunge into details right away, this book can work for you.

I assume you've had a basic algebra background and can do some of the basic mathematical operations and understand some of the basic notation used in algebra like x, y, summation signs, taking the square root, squaring a number, and so on. (If you'd like some backup on the algebra part, I suggest you consider *Algebra I For Dummies* and *Algebra II For Dummies* [Wiley]).

Icons Used in This Book

Here are the road signs you'll encounter on your journey through this book:

Tips refer to helpful hints or shortcuts you can use to save time.

Read these to get the inside track on why a certain concept is important, what its impact will be on the results, and highlights to keep on your radar.

These alert you to common errors that can cause problems, so you can steer around them.

These point out things in the text that you should, if possible, stash away somewhere in your brain for future use.

Where to Go from Here

This book is written in a nonlinear way, so you can start anywhere and still be able to understand what's happening. However, I can make some recommendations for those who are interested in knowing where to start.

For a quick overview of the topics to refresh your memory, check out Chapter 1. For basic number crunching and graphs, see Chapters 2 and 3. If you're most interested in common distributions, see Chapters 4 (binomial); 5 (normal); and 9 (t-distribution). Confidence intervals and hypothesis testing are found in Chapters 7 and 8. Correlation and regression are found in Chapter 10, and two-way tables and independence are tackled in Chapter 11. If you are interested in evaluating and making sense of the results of medical studies, polls, surveys, and experiments, you'll find all the info in Chapters 12 and 13. Common mistakes to avoid or watch for are seen in Chapter 14.

Chapter **1**
Statistics in a Nutshell

The most common description of statistics is that it's the process of analyzing data — number crunching, in a sense. But statistics is not just about analyzing the data. It's about the whole process of using the scientific method to answer questions and make decisions. That process involves designing studies, collecting good data, describing the data with numbers and graphs, analyzing the data, and then making conclusions. In this chapter, I review each of these steps and show where statistics plays the all-important role.

Designing Studies

Once a research question is defined, the next step is designing a study in order to answer that question. This amounts to figuring out what process you'll use to get the data you need. In this section, I overview the two major types of studies: observational studies and experiments.

Surveys

An *observational study* is one in which data are collected on individuals in a way that doesn't affect them. The most common observational study is the survey. *Surveys* are questionnaires that are presented to individuals who have been selected from a population of interest. Surveys take on many different forms: paper surveys sent through the mail; websites; call-in polls conducted by TV networks; and phone surveys. If conducted properly, surveys can be very useful tools for getting information. However, if not conducted properly, surveys can result in bogus information. Some problems include improper wording of questions, which can be misleading, people who were selected to participate but do not respond, or an entire group in the population who had no chance of even being selected. These potential problems mean a survey has to be well thought-out before it's given.

A downside of surveys is that they can only report relationships between variables that are found; they cannot claim cause and effect. For example, if in a survey researchers notice that the people who drink more than one Diet Coke per day tend to sleep fewer hours each night than those who drink at most one per day, they cannot conclude that Diet Coke is causing the lack of sleep. Other variables might explain the relationship, such as number of hours worked per week. See all the information about surveys, their design, and potential problems in Chapter 12.

Experiments

An *experiment* imposes one or more treatments on the participants in such a way that clear comparisons can be made. Once the treatments are applied, the response is recorded. For example, to study the effect of drug dosage on blood pressure, one group might take 10 mg of the drug, and another group might take 20 mg. Typically, a control group is also involved, where subjects each receive a fake treatment (a sugar pill, for example).

Experiments take place in a controlled setting, and are designed to minimize biases that might occur. Some potential problems include: researchers knowing who got what treatment; a certain condition or characteristic wasn't accounted for that can affect the results (such as weight of the subject when studying drug dosage); or lack of a control group. But when designed correctly, if a difference in the responses is found when the groups are compared, the researchers can conclude a cause and effect relationship. See coverage of experiments in Chapter 13.

It is perhaps most important to note that no matter what the study, it has to be designed so that the original questions can be answered in a credible way.

Collecting Data

Once a study has been designed, be it a survey or an experiment, the subjects are chosen and the data are ready to be collected. This phase of the process is also critical to producing good data.

Selecting a good sample

First, a few words about selecting individuals to participate in a study (much, much more is said about this topic in Chapter 12). In statistics, we have a saying: "Garbage in equals garbage out." If you select your subjects in a way that is *biased* — that is, favoring certain individuals or groups of individuals — then your results will also be biased.

Suppose Bob wants to know the opinions of people in your city regarding a proposed casino. Bob goes to the mall with his clipboard and asks people who walk by to give their opinions. What's wrong with that? Well, Bob is only going to get the opinions of a) people who shop at that mall; b) on that particular day; c) at that particular time; d) and who take the time to respond. That's too restrictive — those folks don't represent a cross-section of the city. Similarly, Bob could put up a website survey and ask people to use it to vote. However, only those who know about the site, have Internet access, and want to respond will give him data. Typically, only those with strong opinions will go to such trouble. So, again, these individuals don't represent all the folks in the city.

In order to minimize bias, you need to select your sample of individuals *randomly* — that is, using some type of "draw names out of a hat" process. Scientists use a variety of methods to select individuals at random (more in Chapter 12), but getting a random sample is well worth the extra time and effort to get results that are legitimate.

Avoiding bias in your data

Say you're conducting a phone survey on job satisfaction of Americans. If you call them at home during the day between 9 a.m. and 5 p.m., you'll miss out on all those who work during the day; it could be that day workers are more satisfied than night workers, for example. Some surveys are too long — what if someone stops answering questions halfway through? Or what if they give you misinformation and tell you they make $100,000 a year instead of $45,000? What if they give you an answer that isn't on your list of possible answers? A host of problems can occur when collecting survey data; Chapter 12 gives you tips on avoiding and spotting them.

Experiments are sometimes even more challenging when it comes to collecting data. Suppose you want to test blood pressure; what if the instrument you are using breaks during the experiment? What if someone quits the experiment halfway through? What if something happens during the experiment to distract the subjects or the researchers? Or they can't find a vein when they have to do a blood test exactly one hour after a dose of a drug is given? These are just some of the problems in data collection that can arise with experiments; Chapter 13 helps you find and minimize them.

Describing Data

Once data are collected, the next step is to summarize it all to get a handle on the big picture. Statisticians describe data in two major ways: with pictures (that is, charts and graphs) and with numbers, called *descriptive statistics*.

Descriptive statistics

Data are also summarized (most often in conjunction with charts and/or graphs) by using what statisticians call descriptive statistics. *Descriptive statistics* are numbers that describe a data set in terms of its important features.

If the data are categorical (where individuals are placed into groups, such as gender or political affiliation), they are typically summarized using the number of individuals in each group (called the *frequency*) or the percentage of individuals in each group (the *relative frequency*).

Numerical data represent measurements or counts, where the actual numbers have meaning (such as height and weight). With numerical data, more features can be summarized besides the number or percentage in each group. Some of these features include measures of center (in other words, where is the "middle" of the data?); measures of spread (how diverse or how concentrated are the data around the center?); and, if appropriate, numbers that measure the relationship between two variables (such as height and weight).

Some descriptive statistics are better than others, and some are more appropriate than others in certain situations. For example, if you use codes of 1 and 2 for males and females, respectively, when you go to analyze that data, you wouldn't want to find the average of those numbers — an "average gender" makes no sense. Similarly, using percentages to describe the amount of time until a battery wears out is not appropriate. A host of basic descriptive statistics are presented, compared, and calculated in Chapter 2.

Charts and graphs

Data are summarized in a visual way using charts and/or graphs. Some of the basic graphs used include pie charts and bar charts, which break down variables such as gender and which applications are used on teens' cellphones. A bar graph, for example, may display opinions on an issue using 5 bars labeled in order from "Strongly Disagree" up through "Strongly Agree."

But not all data fit under this umbrella. Some data are numerical, such as height, weight, time, or amount. Data representing counts or measurements need a different type of graph that either keeps track of the numbers themselves or groups them into numerical groupings. One major type of graph that is used to graph numerical data is a histogram. In Chapter 3, you delve into pie charts, bar graphs, histograms and other visual summaries of data.

Analyzing Data

After the data have been collected and described using pictures and numbers, then comes the fun part: navigating through that black box called the *statistical analysis*. If the study has been designed properly, the original questions can be answered using the appropriate analysis, the operative word here being *appropriate*. Many types of analyses exist; choosing the wrong one will lead to wrong results.

In this book, I cover the major types of statistical analyses encountered in introductory statistics. Scenarios involving a fixed number of independent trials where each trial results in either success or failure use the binomial distribution, described in Chapter 4. In the case where the data follow a bell-shaped curve, the normal distribution is used to model the data, covered in Chapter 5.

Chapter 7 deals with confidence intervals, used when you want to make estimates involving one or two population means or proportions using a sample of data. Chapter 8 focuses on testing someone's claim about one or two population means or proportions — these analyses are called hypothesis tests. If your data set is small and follows a bell-shape, the t-distribution might be in order; see Chapter 9.

Chapter 10 examines relationships between two numerical variables (such as height and weight) using correlation and simple linear regression. Chapter 11 studies relationships between two categorical variables (where the data place individuals into groups, such as gender and political affiliation). You can find a fuller treatment of these topics in *Statistics For Dummies* (Wiley), and analyses that are more complex than that are discussed in the book *Statistics II For Dummies,* also published by Wiley.

Making Conclusions

Researchers perform analysis with computers, using formulas. But neither a computer nor a formula knows whether it's being used properly, and they don't warn you when your results are incorrect. At the end of the day, computers and formulas can't tell you what the results mean. It's up to you.

One of the most common mistakes made in conclusions is to overstate the results, or to generalize the results to a larger group than was actually represented by the study. For example, a professor wants to know which Super Bowl commercials viewers liked best. She gathers 100 students from her class on Super Bowl Sunday and asks them to rate each commercial as it is shown. A Top 5 list is formed, and she concludes that Super Bowl viewers liked those five commercials the best. But she really only knows which ones *her students* liked best — she didn't study any other groups, so she can't draw conclusions about all viewers.

Statistics is about much more than numbers. It's important to understand how to make appropriate conclusions from studying data, and that's something I discuss throughout the book.

Chapter **2**
Descriptive Statistics

escriptive statistics are numbers that summarize some characteristic about a set of data. They provide you with easy-to-understand information that helps answer questions. They also help researchers get a rough idea about what's happening in their experiments so later they can do more formal and targeted analyses. Descriptive statistics make a point clearly and concisely.

In this chapter, you see the essentials of calculating and evaluating common descriptive statistics for measuring center and variability in a data set, as well as statistics to measure the relative standing of a particular value within a data set.

Types of Data

Data come in a wide range of formats. For example, a survey might ask questions about gender, race, or political affiliation, while other questions might be about age, income, or the distance you drive to work each day. Different types of questions result in different types of data to be collected and analyzed. The type of data you have determines the type of descriptive statistics that can be found and interpreted.

There are two main types of data: categorical (or qualitative) data and numerical (or quantitative data). *Categorical data* record qualities or characteristics about the individual, such as eye color, gender, political party, or opinion on some issue (using categories such as agree, disagree, or no opinion). *Numerical data* record measurements or counts regarding each individual, which may include weight, age, height, or time to take an exam; counts may include number of pets, or the number of red lights you hit on your way to work. The important difference between the two is that with categorical data, any numbers involved do not have real numerical meaning (for example, using 1 for male and 2 for female), while all numerical data represent actual numbers for which math operations make sense.

TECHNICAL STUFF

A third type of data, *ordinal data*, falls in between, where data appear in categories, but the categories have a meaningful order, such as ratings from 1 to 5, or class ranks of freshman through senior. Ordinal data can be analyzed like categorical data, and the basic numerical data techniques also apply when categories are represented by numbers that have meaning.

Counts and Percents

Categorical data place individuals into groups. For example, male/ female, own your home/don't own, or Democrat/Republican/ Independent/Other. Categorical data often come from survey data, but they can also be collected in experiments. For example, in a test of a new medical treatment, researchers may use three categories to assess the outcome: Did the patient get better, worse, or stay the same?

Categorical data are typically summarized by reporting either the number of individuals falling into each category, or the percentage of individuals falling into each category. For example, pollsters may report the percentage of Republicans, Democrats, Independents, and others who took part in a survey. To calculate the percentage of individuals in a certain category, find the number of individuals in that category, divide by the total number of people in the study, and then multiply by 100%. For example, if a survey of 2,000 teenagers included 1,200 females and 800 males, the resulting percentages would be $(1,200 \div 2,000) * 100\% = 60\%$ female and $(800 \div 2,000) * 100\% = 40\%$ male.

You can further break down categorical data by creating cross-tabs. *Crosstabs* (also called *two-way tables*) are tables with rows and columns. They summarize the information from two categorical variables at once, such as gender and political party, so you can see (or easily calculate) the percentage of individuals in each combination of categories. For example, if you had data about the gender and political party of your respondents, you would be able to look at the percentage of Republican females, Democratic males, and so on. In this example, the total number of possible combinations in your table would be the total number of gender categories times the total number of party affiliation categories. The U.S. government calculates and summarizes loads of categorical data using crosstabs. (see Chapter 11 for more on two-way tables.)

TECHNICAL STUFF

If you're given the number of individuals in each category, you can always calculate your own percents. But if you're only given percentages without the total number in the group, you can never retrieve the original number of individuals in each group. For example, you might hear that 80% of people surveyed prefer Cheesy cheese crackers over Crummy cheese crackers. But how many were surveyed? It could be only 10 people, for all you know, because 8 out of 10 is 80%, just as 800 out of 1,000 is 80%. These two fractions (8 out of 10 and 800 out of 1,000) have different meanings for statisticians, because the first is based on very little data, and the second is based on a lot of data. (See Chapter 7 for more information on data accuracy and margin of error.)

Measures of Center

The most common way to summarize a numerical data set is to describe where the center is. One way of thinking about what the center of a data set means is to ask, "What's a typical value?" Or, "Where is the middle of the data?" The center of a data set can be measured in different ways, and the method chosen can greatly influence the conclusions people make about the data. In this section, I present the two most common measures of center: the mean (or average) and the median.

The *mean* (or average) of a data set is simply the average of all the numbers. Its formula is $\bar{x} = \dfrac{\sum x_i}{n}$. Here is what you need to do to find the mean of a data set, $\bar{x}$:

1. **Add up all the numbers in the data set.**
2. **Divide by the number of numbers in the data set, *n*.**

When it comes to measures of center, the average doesn't always tell the whole story and may be a bit misleading. Take NBA salaries. Every year, a few top-notch players (like Shaq) make much more money than anybody else. These are called *outliers* (numbers in the data set that are extremely high or low compared to the rest). Because of the way the average is calculated, high outliers drive the average upward (as Shaq's salary did in the preceding example). Similarly, outliers that are extremely low tend to drive the average downward.

What can you report, other than the average, to show what the salary of a "typical" NBA player would be? Another statistic used to measure the center of a data set is the median. The *median* of a data set is the place that divides the data in half, once the data are ordered from smallest to largest. It is denoted by M or $\tilde{x}$. To find the median of a data set:

1. **Order the numbers from smallest to largest.**
2. **If the data set contains an odd number of numbers, the one exactly in the middle is the median.**
3. **If the data set contains an even number of numbers, take the two numbers that appear exactly in the middle and average them to find the median.**

For example, take the data set 4, 2, 3, 1. First, order the numbers to get 1, 2, 3, 4. Then note this data has an even number of numbers, so go to Step 3. Take the two numbers in the middle — 2 and 3 — and find their average: 2.5.

Note that if the data set is odd, the median will be one of the numbers in the data set itself. However, if the data set is even, it may be one of the numbers (the data set 1, 2, 2, 3 has median 2); or it may not be, as the data set 4, 2, 3, 1 (whose median is 2.5) shows.

Which measure of center should you use, the mean or the median? It depends on the situation, but reporting both is never a bad idea. Suppose you're part of an NBA team trying to negotiate salaries. If you represent the owners, you want to show how much everyone is making and how much you're spending, so you want to take into account those superstar players and report the average. But if you're on the side of the players, you want to report the median, because that's more representative of what the players in the middle are making. Fifty percent of the players make a salary above the median, and 50% make a salary below the median.

REMEMBER

When the mean and median are not close to each other in terms of their value, it's a good idea to report both and let the reader interpret the results from there. Also, as a general rule, be sure to ask for the median if you are only given the mean.

Measures of Variability

Variability is what the field of statistics is all about. Results vary from individual to individual, from group to group, from city to city, from moment to moment. Variation always exists in a data set, regardless of which characteristic you're measuring, because not every individual will have the same exact value for every characteristic you measure. Without a measure of variability you can't compare two data sets effectively. What if in both sets two sets of data have about the same average and the same median? Does that mean that the data are all the same? Not at all. For example, the data sets 199, 200, 201, and 0, 200, 400 both have the same average, which is 200, and the same median, which is also 200. Yet they have very different amounts of variability. The first data set has a very small amount of variability compared to the second.

By far the most commonly used measure of variability is the standard deviation. The *standard deviation* of a data set, denoted by s, represents the typical distance from any point in the data set to the center. It's roughly the average distance from the center, and in this case, the center is the average. Most often, you don't hear a standard deviation given just by itself; if it's reported (and it's not reported nearly enough) it's usually in the fine print, in parentheses, like "($s = 2.68$)."

The formula for the standard deviation of a data set is $s = \sqrt{\dfrac{\sum(x-\bar{x})^2}{n-1}}$. To calculate s, do the following steps:

1. **Find the average of the data set, $\bar{x}$.** To find the average, add up all the numbers and divide by the number of numbers in the data set, n.

2. **For each number, subtract the average from it.**

3. **Square each of the differences.**

4. **Add up all the results from Step 3.**

5. **Divide the sum of squares (Step 4) by the number of numbers in the data set, minus one ($n-1$).**

   If you do Steps 1 through 5 only, you have found another measure of variability, called the *variance*.

6. **Take the square root of the variance. This is the standard deviation.**

   Suppose you have four numbers: 1, 3, 5, and 7. The mean is $16 \div 4 = 4$. Subtracting the mean from each number, you get $(1-4) = -3, (3-4) = -1, (5-4) = +1$, and $(7-4) = +3$. Squaring the results you get 9, 1, 1, and 9, which sum to 20. Divide 20 by $4 - 1 = 3$ to get 6.67. The standard deviation is the square root of 6.67, which is 2.58.

Here are some properties that can help you when interpreting a standard deviation:

» The standard deviation can never be a negative number.

» The smallest possible value for the standard deviation is 0 (when every number in the data set is exactly the same).

» Standard deviation is affected by outliers, as it's based on distance from the mean, which is affected by outliers.

» The standard deviation has the same units as the original data, while variance is in square units.

Percentiles

The most common way to report relative standing of a number within a data set is by using *percentiles*. A percentile is the percentage of individuals in the data set who are below where your particular number is located. If your exam score is at the 90th percentile, for example, that means 90% of the people taking the exam with you scored lower than you did (it also means that 10 percent scored higher than you did.)

Finding a percentile

To calculate the kth percentile (where k is any number between one and one hundred), do the following steps:

1. Order all the numbers in the data set from smallest to largest.

2. Multiply k percent times the total number of numbers, n.

3a. If your result from Step 2 is a whole number, go to Step 4. If the result from Step 2 is not a whole number, round it up to the nearest whole number and go to Step 3b.

3b. Count the numbers in your data set from left to right (from the smallest to the largest number) until you reach the value from Step 3a. This corresponding number in your data set is the kth percentile.

4. Count the numbers in your data set from left to right until you reach that whole number. The kth percentile is the average of that corresponding number in your data set and the next number in your data set.

For example, suppose you have 25 test scores, in order from lowest to highest: 43, 54, 56, 61, 62, 66, 68, 69, 69, 70, 71, 72, 77, 78, 79, 85, 87, 88, 89, 93, 95, 96, 98, 99, 99. To find the 90th percentile for these (ordered) scores start by multiplying 90% times the total number of scores, which gives $90\% \times 25 = 0.90 \times 25 = 22.5$ (Step 2). This is not a whole number; Step 3a says round up to the nearest whole number — 23 — then go to Step 3b. Counting from left to right (from the smallest to the largest number in the data set), you go until you find the 23rd number in the data set. That number is 98, and it's the 90th percentile for this data set.

If you want to find the 20th percentile, take $0.20 * 25 = 5$; this is a whole number so proceed to Step 4, which tells you the 20th percentile is the average of the 5th and 6th numbers in the ordered data set (62 and 66). The 20th percentile then comes to $\frac{(62+66)}{2} = 64$.

TIP

The median is the 50th percentile, the point in the data where 50% of the data fall below that point and 50% fall above it. The median for the test scores example is the 13th number, 77.

Interpreting percentiles

The U.S. government often reports percentiles among its data summaries. For example, the U.S. Census Bureau reported the median household income for 2001 was $42,228. The Bureau also reported various percentiles for household income, including the 10th, 20th, 50th, 80th, 90th, and 95th. Table 2-1 shows the values of each of these percentiles.

TABLE 2-1 ## U.S. Household Income for 2001

Percentile	2001 Household Income
10th	$10,913
20th	$17,970
50th	$42,228
80th	$83,500
90th	$116,105
95th	$150,499

Looking at these percentiles, you can see that the bottom half of the incomes are closer together than are the top half. The difference between the 50th percentile and the 20th percentile is about $24,000, whereas the spread between the 50th percentile and the 80th percentile is more like $41,000. And the difference between the 10th and 50th percentiles is only about $31,000, whereas the difference between the 90th and the 50th percentiles is a whopping $74,000.

TECHNICAL STUFF

A percentile is *not* a percent; a percentile is a number that is a certain percentage of the way through the data set, when the data set is ordered. Suppose your score on the GRE was reported to be the 80th percentile. This doesn't mean you scored 80% of the questions correctly. It means that 80% of the students' scores were lower than yours, and 20% of the students' scores were higher than yours.

The Five-Number Summary

The five-number summary is a set of five descriptive statistics that divide the data set into four equal sections. The five numbers in a five number summary are:

1. The minimum (smallest) number in the data set

2. The 25th percentile, aka the first quartile, or Q1

3. The median (or 50th percentile)

4. The 75th percentile, aka the third quartile, or Q3

5. The maximum (largest) number in the data set

For example, you can find the five-number summary of the 25 (ordered) exam scores 43, 54, 56, 61, 62, 66, 68, 69, 69, 70, 71, 72, 77, 78, 79, 85, 87, 88, 89, 93, 95, 96, 98, 99, 99. The minimum is 43, the maximum is 99, and the median is the number directly in the middle, 77.

To find Q1 and Q3, you use the steps shown in the section "Finding a percentile," where $n = 25$. Step 1 is done since the data are ordered. For Step 2, since Q1 is the 25th percentile, multiply $0.25 * 25 = 6.25$. This is not a whole number, so Step 3a says round it up to 7 and proceed to Step 3b. Count from left to right in the data set until you reach the 7th number, 68; this is Q1. For Q3 (the 75th percentile), multiply $0.75 * 25 = 18.75$; round up to 19, and the 19th number on the list is 89, or Q3. Putting it all together, the five-number summary for the test scores data is 43, 68, 77, 89, and 99.

The purpose of the five-number summary is to give descriptive statistics for center, variability, and relative standing all in one shot. The measure of center in the five-number summary is the median, and the first quartile, median, and third quartiles are measures of relative standing. To obtain a measure of variability based on the five-number summary, you can find what's called the *Interquartile Range* (IQR). The IQR equals $Q3 - Q1$ and reflects the distance taken up by the innermost 50% of the data. If the IQR is small, you know there is much data close to the median. If the IQR is large, you know the data are more spread out from the median. The IQR for the test scores data set is $89 - 68 = 21$, which is quite large seeing as how test scores only go from 0 to 100.

Chapter **3**

Charts and Graphs

The main purpose of a data display is to organize and display data to make your point clearly, effectively, and correctly. In this chapter, I present the most common data displays used to summarize categorical and numerical data, thoughts and cautions on their interpretation, and tips for evaluating them.

Pie Charts

A pie chart takes categorical data and shows the percentage of individuals that fall into each category. The sum of all the slices of the pie should be 100% or close to it (with a bit of round-off error). Because a pie chart is a circle, categories can easily be compared and contrasted to one another.

The Florida lottery uses a pie chart to report where your money goes when you purchase a lottery ticket (see Figure 3-1). You can see that half of Florida lottery revenues (50 cents of every dollar spent) goes to prizes, and 38 cents of every dollar goes to education.

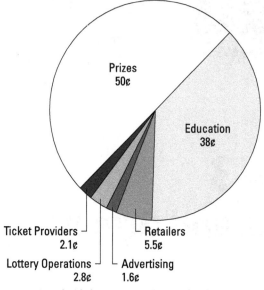

FIGURE 3-1: Florida lottery expenditures (fiscal year 2001–2002).

To evaluate a pie chart for statistical correctness:

» Check to be sure the percentages add up to 100% or close to it (any round-off error should be very small).

» Beware of slices of the pie called "other" that are larger than many of the other slices. This shows a lack of detail in the information gathered.

» A pie chart only shows the percentage in each group, not the number in each group. Always ask for or look for a report of the total size of the data set.

Bar Graphs

A bar graph is another means for summarizing categorical data. Like a pie chart, a bar graph breaks categorical data down by group, showing how many individuals lie in each group, or what percentage lies in each group.

Bar graphs are often used to compare groups by breaking down the categories for each and showing them as side-by-side bars. For example, has the percentage of mothers in the workforce changed over time? Figure 3-2 says yes and shows that the overall percentage of mothers in the workforce climbed from 47% to 72% between 1975 and 1998. Taking the age of the child into account, fewer mothers work while their children are younger and not in school yet, but the difference from 1975 to 1998 is still about 25% in each case.

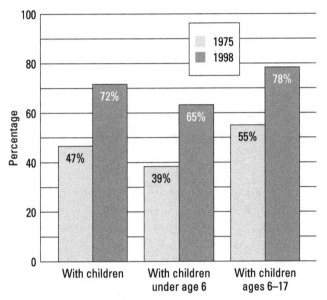

FIGURE 3-2: Percentage of mothers in workforce, by age of child (1975 and 1998 — data are from the U.S. Census).

Here is a checklist for evaluating bar graphs:

» Check the units on the y-axis. Make sure they are evenly spaced.

» Be aware of the scale of the bar graph (the units in which bar heights are represented). Using a smaller scale (for example, each half inch of height representing 10 units versus 50), you can make differences look more dramatic.

» In the case where the bars represent percents and not counts, make sure to ask for the total number of individuals summarized by the bar graph if it is not listed.

Time Charts

A *time chart* is a data display whose main point is to examine trends over time. Another name for a time chart is a *line graph*. Typically a time chart has some unit of time on the horizontal axis (year, day, month, and so on) and a measured quantity on the vertical axis (average household income, birth rate, total sales, and so on). At each time period, the amount is shown as a dot, and the dots connect to form the time chart.

You can see in Figure 3-3 that wages for production workers, when adjusted for inflation, increased from 1947 until the early 1970s, declined during the 1970s, and basically stayed in the same range until the late 1990s, when a small surge began.

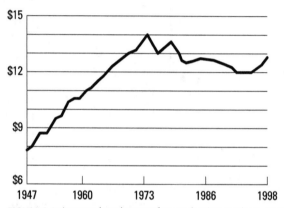

FIGURE 3-3: Average hourly wage for production workers, 1947–1998 (in 1998 dollars).

TECHNICAL STUFF

A time chart can present information in a misleading way, such as charting the *number* of crimes over time, rather than the crime *rate* (crimes per capita). Because the population size of a city changes over time, crime rate is the appropriate measure. Make sure you understand what statistics are being presented and examine them for fairness and appropriateness.

Here is a checklist for evaluating time charts:

>> Examine the scale on the vertical (quantity) axis as well as the horizontal (timeline) axis; results can be made to look more or less dramatic than they actually are simply by changing the scale.

>> Take into account the units used in the chart and be sure they're appropriate for comparison over time (for example, are dollar amounts adjusted for inflation?).

>> Watch for gaps in the timeline on a time chart. Connecting the dots across a short period of time is better than connecting across a long time.

Histograms

A histogram is the statistician's graph of choice for numerical data. It provides a snapshot of all the data broken down into numerically ordered groups. Histograms provide a quick way to get the big idea about a numerical data set.

Making a histogram

A *histogram* is basically a bar graph that applies to numerical data. Because the data are numerical, the categories are ordered from smallest to largest (as opposed to categorical data, such as gender, which has no inherent order to it). To be sure each number falls into exactly one group, the bars on a histogram touch each other but don't overlap. Each bar is marked on the x-axis (horizontal) by the values representing its beginning and endpoints. The height of each bar of a histogram represents either the number of individuals in each group (the *frequency* of each group) or the percentage of individuals in each group (the *relative frequency* of each group).

Table 3-1 shows the number of live births in Colorado by age of mother for selected years from 1975–2000. The numerical variable age is broken down into categories of 5-year groupings. Relative frequency histograms comparing 1975 and 2000 are shown in Figure 3-4. You can see more older mothers in 2000 than in 1975.

TECHNICAL STUFF

If a data point falls directly on a borderline between two groups, be consistent in deciding which group to place that value into. For example, if the groups are 0–5, 5–10, 10–15, and you get a data point of 10, you can include it either in the 5–10 group or the 10–15 group, as long as you are consistent with other data falling on borderlines.

TABLE 3-1 Colorado Live Births by Mother's Age

Year	Total births	10–14	15–19	20–24	25–29	30–34	35–39	40–44	45–49
1975	40,148	88	6,627	14,533	12565	4,885	1,211	222	16
1980	49,716	57	6,530	16,642	16,081	8,349	1,842	198	12
1985	55,115	90	5,634	16,242	18,065	11,231	3,464	370	13
1990	53,491	91	5,975	13,118	16,352	12,444	4,772	717	15
1995	54,310	134	6,462	12,935	14,286	13,186	6,184	1,071	38
2000	65,429	117	7,546	15,865	17,408	15,275	7,546	1,545	93

Note: The sum of births may not add up to the total number of births due to unknown or unusually high age (50 and over) of the mother.

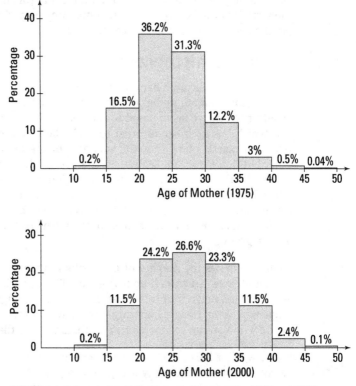

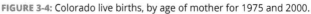
FIGURE 3-4: Colorado live births, by age of mother for 1975 and 2000.

Interpreting a histogram

A histogram tells you three main features of numerical data:

» How the data are distributed (symmetric, skewed right, skewed left, bell-shaped, and so on)

» The amount of variability in the data

» Where the center of the data is (approximately)

The distribution of the data in a histogram

One of the features that a histogram can show you is the so-called *shape* of the data (in other words, how the data are distributed among the groups). Many shapes exist, and many data sets show a combination of shapes, but there are three major shapes to look for in a data set:

1. *Symmetric,* meaning that the left-hand side of the histogram is a mirror image of the right-hand side

2. *Skewed right,* meaning that it looks like a lopsided mound with one long tail going off to the right

3. *Skewed left,* meaning that it looks like a lopsided mound with one long tail going off to the left

Mothers' ages in Figure 3-4 for years 1975 and 2000 appear to be mostly mound-shaped, although the data for 1975 are slightly skewed to the right, indicating that as women got older, fewer had babies relative to the situation in 2000. In other words, in 2000 a higher proportion of older women were having babies compared to 1975.

Variability in the data from a histogram

You can also get a sense of variability in the data by looking at a histogram. If a histogram is quite flat with the bars close to the same height, you may think it indicates less variability, but in fact the opposite is true. That's because you have an equal number in each bar, but the bars themselves represent different ranges of values, so the entire data set is actually quite spread out. A histogram with a big lump in the middle and tails on the sides indicates more data in the middle bars than the outer bars, so the data are actually closer together.

Comparing 1975 to 2000, there's more variability in 2000. This, again, indicates changing times; more women are waiting to have children (in 1975 most women had their children by age 30), and the length of time waiting varies. (Chapter 2 discusses measuring variability in a data set.)

TECHNICAL STUFF

Variability in a histogram should not be confused with variability in a time chart. If values change over time, they're shown on a time chart as highs and lows, and many changes from high to low (over time) indicate lots of variability. So, a flat line on a time chart indicates no change and no variability in the values across time. But when the heights of histogram bars appear flat (uniform), this shows values spread out uniformly over many groups, indicating a great deal of variability in the data at one point in time.

Center of the data from a histogram

A histogram can also give you a rough idea of where the center of the data lies. To visualize the mean, picture the data as people on a teeter-totter; the mean is the point where the fulcrum has to be in order to balance the weight on each side.

Note in Figure 3-4 that the mean appears to be around 25 years for 1975 and around 27.5 years for 2000. This suggests that in 2000, Colorado women were having children at older ages, on average, than they did in 1975.

Evaluating a histogram

Here is a checklist for evaluating a histogram:

>> Examine the scale used for the vertical (frequency or relative frequency) axis and beware of results that appear exaggerated or played down through the use of inappropriate scales.

>> Check out the units on the vertical axis to see whether the histogram reports frequencies (numbers) or relative frequencies (percentages), and then take this into account when evaluating the information.

>> Look at the scale used for the groupings of the numerical variable (on the horizontal axis). If the range for each group is very small, the data may look overly volatile. If the ranges are very large, the data may appear to be smoother than they really are.

Boxplots

A *boxplot* is a one-dimensional graph of numerical data based on the five-number summary, which includes the minimum value, the 25th percentile (known as Q_1), the median, the 75th percentile (Q_3), and the maximum value. In essence, these five descriptive statistics divide the data set into four equal parts. (See Chapter 2 for more on the five-number summary.)

Making a boxplot

To make a boxplot, follow these steps:

1. Find the five-number summary of your data set. (Use the steps outlined in Chapter 2.)

2. Create a horizontal number line whose scale includes the numbers in the five-number summary.

3. Label the number line using appropriate units of equal distance from each other.

4. Mark the location of each number in the five-number summary just above the number line.

5. Draw a box around the marks for the 25th percentile and the 75th percentile.

6. Draw a line in the box where the median is located.

7. Draw lines from the outside edges of the box out to the minimum and maximum values in the data set.

Consider the following 25 exam scores: 43, 54, 56, 61, 62, 66, 68, 69, 69, 70, 71, 72, 77, 78, 79, 85, 87, 88, 89, 93, 95, 96, 98, 99, and 99. The five-number summary for these exam scores is 43, 68, 77, 89, and 99, respectively. (This data set is described in detail in Chapter 2.) The vertical version of the boxplot for these exam scores is shown in Figure 3-5.

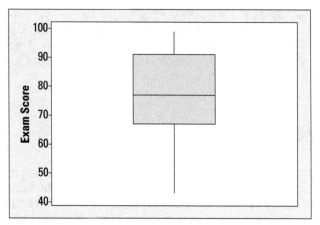

FIGURE 3-5: Boxplot of 25 exam scores.

TECHNICAL
STUFF

Some statistical software adds asterisk signs (*) to show numbers in the data set that are considered to be *outliers* — numbers determined to be far enough away from the rest of the data to be noteworthy.

Interpreting a boxplot

A boxplot can show information about the distribution, variability, and center of a data set.

Distribution of data in a boxplot

A boxplot can show whether a data set is symmetric (roughly the same on each side when cut down the middle), or skewed (lopsided). Symmetric data show a symmetric boxplot; skewed data show a lopsided boxplot, where the median cuts the box into two unequal pieces. If the longer part of the box is to the right (or above) the median, the data are said to be *skewed right*. If the longer part is to the left (or below) the median, the data are *skewed left*. However, no data set falls perfectly into one category or the other.

In Figure 3-5, the upper part of the box is wider than the lower part. This means that the data between the median (77) and Q_3 (89) are a little more spread out, or variable, than the data between the median (77) and Q_1 (68). You can also see this by subtracting $89 - 77 = 12$ and comparing to $77 - 68 = 9$. This indicates the data in the middle 50% of the data set are a bit skewed right. However, the line between the min (43) and Q_1 (68) is longer than the line between Q_3 (89) and the max (99). This indicates a "tail" in the data trailing to the left; the low exam scores are spread out quite a bit more than the high ones. This greater difference causes the overall shape of the data to be skewed left. (Since there are no strong outliers on the low end, you can safely say that the long tail is not due to an outlier.) A histogram of the exam data, shown in the graph in Figure 3-6, confirms the data are generally skewed left.

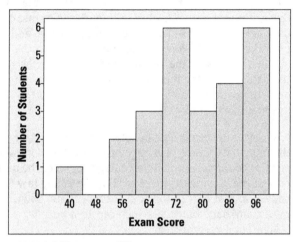

FIGURE 3-6: Histogram of 25 exam scores.

TECHNICAL STUFF

A boxplot can tell you whether a data set is symmetric, but it can't tell you the shape of the symmetry. For example, a data set like 1, 1, 2, 2, 3, 3, 4, 4 is symmetric and each number appears the same number of times, whereas 1, 2, 2, 2, 3, 4, 5, 5, 5, 6 is also symmetric but doesn't have an equal number of values in each group. Boxplots of both would look similar in shape. A histogram shows the particular shape that the symmetry has.

Variability in a data set from a boxplot

Variability in a data set that is described by the five-number summary is measured by the *interquartile range* (IQR — see Chapter 2 for full details on the IQR). The interquartile range is equal to $Q_3 - Q_1$. A large distance from the 25th percentile to the 75th indicates the data are more variable. Notice that the IQR ignores data below the 25th percentile or above the 75th, which may contain outliers that could inflate the measure of variability of the entire data set. In the exam score data, the IQR is $89 - 68 = 21$, compared to the range of the entire data set (max− min = 56). This indicates a fairly large spread within the innermost 50% of the exam scores.

Center of the data from a boxplot

The median is part of the five-number summary, and is shown by the line that cuts through the box in the boxplot. This makes it very easy to identify. The mean, however, is not part of the boxplot, and couldn't be determined accurately from a boxplot. In the exam score data, the median is 77. Separate calculations show the mean to be 76.96. These are extremely close, and my reasoning is because the skewness to the right within the middle 50% of the data offsets the skewness to the left of the outer part of the data. To get the big picture of any data set you need to find more than one measure of center and spread, and show more than one graph, as the ideal report.

It's easy to misinterpret a boxplot by thinking the bigger the box, the more data. Remember each of the four sections shown in the boxplot contains an equal percentage (25%) of the data. A bigger part of the box means there is more *variability* (a wider range of values) in that part of the box, not more data. You can't even tell how many data values are included in a boxplot — it is totally built around percentages.

Chapter **4**

The Binomial Distribution

A random variable is a characteristic, measurement, or count that changes randomly according to some set of probabilities; its notation is X, Y, Z, and so on. A list of all possible values of a random variable, along with their probabilities, is called a probability distribution. One of the most well-known probability distributions is the binomial. Binomial means "two names" and is associated with situations involving two outcomes: success or failure (hitting a red light or not; developing a side effect or not). This chapter focuses on the binomial distribution — when you can use it, finding probabilities for it, and finding the expected value and variance.

Characteristics of a Binomial

A random variable has a binomial distribution if all the following conditions are met:

1. There are a fixed number of trials (n).
2. Each trial has two possible outcomes: success or failure.

3. The probability of success (call it p) is the same for each trial.

4. The trials are independent, meaning the outcome of one trial doesn't influence that of any other.

Let X equal the total number of successes in n trials; if all the above conditions are met, X has a binomial distribution with probability of success equal to p.

Checking the binomial conditions step by step

You flip a fair coin 10 times and count the number of heads. Does this represent a binomial random variable? You can check by reviewing your responses to the questions and statements in the list that follows:

1. **Are there a fixed number of trials?**

You're flipping the coin 10 times, which is a fixed number. Condition 1 is met, and $n = 10$.

2. **Does each trial have only two possible outcomes — success or failure?**

The outcome of each flip is either heads or tails, and you're interested in counting the number of heads, so flipping a head represents success and flipping a tail is a failure. Condition 2 is met.

3. **Is the probability of success the same for each trial?**

Because the coin is fair, the probability of success (getting a head) is $p = \frac{1}{2}$ for each trial. You also know that $1 - \frac{1}{2} = \frac{1}{2}$ is the probability of failure (getting a tail) on each trial. Condition 3 is met.

4. **Are the trials independent?**

We assume the coin is being flipped the same way each time, which means the outcome of one flip doesn't affect the outcome of subsequent flips. Condition 4 is met.

Non-binomial examples

Because the coin-flipping example meets the four conditions, the random variable X, which counts the number of successes (heads) that occur in 10 trials, has a binomial distribution with $n = 10$ and $p = \frac{1}{2}$. But not every situation that appears binomial actually is binomial. Consider the following examples.

No fixed number of trials

Suppose now you are to flip a fair coin until you get four heads, and you count how many flips it takes to get there. (That is, X is the number of flips needed.) This certainly sounds like a binomial situation: Condition 2 is met since you have success (heads) and failure (tails) on each flip; Condition 3 is met with the probability of success (heads) being the same (0.5) on each flip; and the flips are independent, so Condition 4 is met.

However, notice that X isn't counting the number of heads, it counts the number of trials needed to get four heads. The number of successes (X) is fixed rather than the number of trials (n). Condition 1 is not met, so X does not have a binomial distribution in this case.

More than success or failure

Some situations involve more than two possible outcomes yet they can appear to be binomial. For example, suppose you roll a fair die 10 times and record the outcome each time. You have a series of $n = 10$ trials, they are independent, and the probability of each outcome is the same for each roll. However, you're recording the outcome on a six-sided die. This is not a success/failure situation, so Condition 2 is not met.

However, depending on what you're recording, situations originally having more than two outcomes can fall under the binomial category. For example, if you roll a fair die 10 times and each time record whether or not you get a 1, then Condition 2 is met because your two outcomes of interest are getting a 1 ("success") and not getting a 1 ("failure"). In this case $p = 1/6$ is the probability for a success and 5/6 for failure. This is a binomial.

Probability of success (p) changes

You have 10 people — 6 women and 4 men — and form a committee of 2 at random. You choose a woman first with probability 6/10. The chance of selecting another woman is now 5/9. The value of p has changed, and Condition 3 is not met. This happens with small populations where replacing an individual after he or she is chosen (to keep probabilities the same) doesn't make sense. You can't choose someone twice for a committee.

Trials are not independent

The independence condition is violated when the outcome of one trial affects another trial. Suppose you want to know support levels of adults in your city for a proposed casino. Instead of taking a random sample of say 100 people, to save time you select 50 married couples and ask each individual what his or her opinion is. Married couples have a higher chance of agreeing on their opinions than individuals selected at random, so the independence Condition 4 is not met.

Finding Binomial Probabilities Using the Formula

After you identify that X has a binomial distribution (the four conditions are met), you'll likely want to find probabilities for X. The good news is that you don't have to find them from scratch; you get to use previously established formulas for finding binomial probabilities, using the values of n and p unique to each problem.

Probabilities for a binomial random variable X can be found using the formula $\binom{n}{x} p^{x} \left(1-p\right)^{n-x}$, where

>> n is the fixed number of trials.

>> x is the specified number of successes.

>> $n-x$ is the number of failures.

>> p is the probability of success on any given trial.

>> $1-p$ is the probability of failure on any given trial. (***Note:*** Some textbooks use the letter q to denote the probability of failure rather than $1-p$.)

These probabilities hold for any value of X between 0 (lowest number of possible successes in n trials) and n (highest number of possible successes).

TECHNICAL STUFF

The number of ways to arrange x successes among n trials is called "n choose x," and the notation is $\binom{n}{x}$. For example, $\binom{3}{2}$ means "3 choose 2" and stands for the number of ways to get 2 successes in 3 trials. In general, to calculate "n choose x," you use the formula

$$\binom{n}{x} = \frac{n!}{x!(n-x)!}.$$ The notation $n!$ stands for n-*factorial*, the number of ways to rearrange n items. To calculate $n!$, you multiply $n(n-1)(n-2)...(2)(1)$. For example 3! is $3(2)(1) = 6$; 2! is $2(1) = 2$; and 1! is 1. By convention, 0! equals 1. To calculate "3 choose 2," you do the following:

$$\binom{3}{2} = \frac{3!}{2!(3-2)!} = \frac{3*2*1}{(2*1)(1!)} = \frac{6}{2*1} = 3$$

Suppose you cross three traffic lights on your way to work, and the probability of each of them being red is 0.30. (Assume the lights are independent.) You let X be the number of red lights you encounter and you want to find the probability distribution for X. You know $p =$ probability of red light $= 0.30$; $1 - p =$ probability of a non-red light $= 1 - 0.30 = 0.70$; and the number of non-red lights is $3 - X$. Using the formula, you obtain the probabilities for $X = 0, 1, 2,$ and 3 red lights:

$$P(X=0) = \binom{3}{0}0.30^0(1-0.30)^{3-0} = \frac{3!}{0!(3-0)!}(0.30)^0(0.70)^3 =$$
$$1(1)(0.343) = 0.343$$

$$P(X=1) = \binom{3}{1}0.30^1(1-0.30)^{3-1} = \frac{3!}{1!(3-1)!}(0.30)^1(0.70)^2 =$$
$$3(0.30)(0.49) = 0.441$$

$$P(X=2) = \binom{3}{2}0.30^2(1-0.30)^{3-2} = \frac{3!}{2!(3-2)!}(0.30)^2(0.70)^1 =$$
$$3(0.09)(0.70) = 0.189$$

$$P(X=3) = \binom{3}{3}.30^3(1-.30)^{3-3} = \frac{3!}{3!(3-3)!}(0.30)^3(0.70)^0 =$$
$$1(0.027)(1) = 0.027$$

The final probability distribution for X is shown in Table 4-1. Notice they sum to 1 because every possible value of X is listed and accounted for.

TABLE 4-1 **Probability Distribution for X = Number of Red Traffic Lights ($n = 3, p = 0.30$)**

X	p(x)
0	0.343
1	0.441
2	0.189
3	0.027

Finding Probabilities Using the Binomial Table

A large range of binomial probabilities are already provided in Table A-3 in the appendix (called the binomial table). In Table A-3 you see several mini-tables provided in the binomial table; each one corresponds with a different n for a binomial (various values of n up to 20 are available). Each mini-table has rows and columns. Running down the side of any mini-table, you see all the possible values of X from 0 through n, each with its own row. The columns of Table A-3 represent various values of p up through and including 0.50. (When $p > 0.50$, a slight change is needed to use Table A-3, as I explain later in this section.)

Finding probabilities when p ≤ 0.50

To use Table A-3 (in the appendix) to find binomial probabilities for X when $p < 0.50$, follow these steps:

1. **Find the mini-table associated with your particular value of *n* (the number of trials).**

2. **Find the column that represents your particular value of *p* (or the one closest to it).**

3. **Find the row that represents the number of successes (*x*) you are interested in.**

4. **Intersect the row and column from Steps 2 and 3 in Table A-3.** This gives you the probability for *x* successes.

For the traffic light example, you can use Table A-3 (appendix) to verify the results found by the binomial formula shown in Table 4-1 (previous section). In Table A-3, go to the mini-table where $n = 3$, and look in the column where $p = 0.30$. You see four probabilities listed for this mini-table: 0.3430; 0.4410; 0.1890; and 0.0270; these are the probabilities for $X = 0$, 1, 2, and 3 red lights, respectively, matching those from Table 4-1.

Finding probabilities when p > 0.50

Notice that Table A-3 (appendix) shows binomial probabilities for several different values of n and p, but the values of p only go up through 0.50. This is because it's still possible to use Table A-3 to find probabilities when p is greater than 0.50. You do it by counting failures (whose probabilities are $1 - p$) instead of successes. When $p \geq 0.50$, you know $(1 - p) < 0.50$.

To use the Table A-3 to find probabilities for X when $p > 0.50$, follow these steps:

1. **Find the mini-table associated with your particular value of *n* (the number of trials).**

2. **Instead of looking at the column for the probability of success (*p*), find the column that represents $1 - p$, the probability of a failure.**

3. **Find the row that represents the number of failures ($n - x$) that are associated with the number of successes (*x*) you want.**

   For example, if you want the chance of 3 successes in 10 trials, it's the same as the chance of 7 failures, so look in row 7.

4. **Intersect the row and column from Steps 2 and 3 in Table A-3 and you see the probability for the number of failures you counted.**

   This also equals the probability for the number of successes (*x*) that you wanted.

Once you've done Step 4, you're done. You do not need to take the complement of your final answer. The complements were taken care of by using the $1 - p$ and counting failures instead of successes.

Revisiting the traffic light example, suppose you are now driving on side streets in your city and you still have 3 intersections ($n = 3$) but now the chance of a red light is $p = 0.70$. Again, let X represent the number of red lights. Table A-3 has no column for $p = 0.70$. However, if the probability of a red light is $p = 0.70$, then the probability of a non-red light $1 - .70 = 0.30$; so instead of counting red lights, you count non-red lights.

Let Y count the number of non-red lights in the three intersections; Y is binomial with $n = 3$ and $p = 0.30$. The probability distribution for Y is shown in Table 4-2. This is also the probability distribution for X, the number of red lights ($n = 3$ and $p = 0.70$), which is what you originally asked for.

TABLE 4-2 Probability Distribution for the Number of Red Traffic Lights ($n = 3, p = 0.70$)

X = number of red	Y = number of non-red	Probability
0	3	0.027
1	2	0.189
2	1	0.441
3	0	0.343

Finding probabilities for X greater-than, less-than, or between two values

Table A-3 (appendix) shows probabilities for X being equal to any value from 0 to n, for a variety of ps. To find probabilities for X being less-than, greater-than, or between two values, just find the corresponding values in the table and add their probabilities. For the traffic light example where $n = 3$ and $p = 0.70$, if you want $P(X > 1)$, you find $P(X = 2) + P(X = 3)$ and get $0.441 + 0.343 = 0.784$. The probability that X is between 1 and 3 (inclusive) is $0.189 + 0.441 + 0.343 = 0.973$.

TIP

Two phrases to remember: "at-least" means that number or higher; "at-most" means that number or lower. For example the probability that X is at least 2 is $P(X \geq 2)$; the probability that X is at most 2 is $P(X \leq 2)$.

The Expected Value and Variance of the Binomial

The mean of a random variable is the long-term average of its possible values over the entire population of individuals (or trials). It's found by taking the weighted average of the x-values multiplied by their probabilities. The mean of a random variable is denoted by μ. For the binomial random variable the mean is $\mu = np$.

Suppose you flip a fair coin 100 times and let X be the number of heads; this is a binomial random variable with $n = 100$ and $p = 0.50$. Its mean is $np = 100(0.50) = 50$.

The variance of a random variable X is the weighted average of the squared deviations (distances) from the mean. The variance of a random variable is denoted by σ^2. The variance of the binomial distribution is $\sigma^2 = np(1-p)$. The standard deviation of X is just the square root of the variance, which in this case is $\sigma^2 = \sqrt{np(1-p)}$.

Suppose you flip a fair coin 100 times and let X be the number of heads. The variance of X is $(1-p) = 100(0.50)(1-0.50) = 25$, and the standard deviation is the square root, which is 5.

TIP

The mean and variance of a binomial have intuitive meaning. The p is the probability of a success, but it also represents the **proportion** of successes you can expect in n trials. Therefore the total **number** of successes you can expect — that is, the mean of X — equals np. The only variability in the outcomes of each trial is between success (with probability p) and failure (with probability $1-p$). Over n trials, it makes sense that the variance of the number of successes/failures is measured by $np(1-p)$.

Chapter **5**
The Normal Distribution

There are two major types of random variables: discrete and continuous. Discrete random variables basically count things (number of heads on 10 coin flips, number of female Democrats in a sample, and so on). The most well known discrete random variable is the binomial (see Chapter 4). A continuous random variable measures things and takes on values within an interval, or they have so many possible values that they might as well be deemed continuous (for example, time to complete a task, exam scores, and so on).

In this chapter, you work on finding probabilities for the most famous continuous random variable, the normal. You also find percentiles for the normal distribution (where you are given a probability as a percent) and you have to find the value of X that's associated with it.

Basics of the Normal Distribution

We say that X has a normal distribution if its values fall into a smooth (continuous) curve with a bell-shaped, symmetric pattern, meaning it looks the same on each side when cut down the middle. The total area under the curve is 1. Each normal distribution has its own mean, μ, and its own standard deviation, σ.

For intro stat courses, the mean and standard deviation for the normal distribution are given to you.

Figure 5-1 illustrates three different normal distributions with different means and standard deviations.

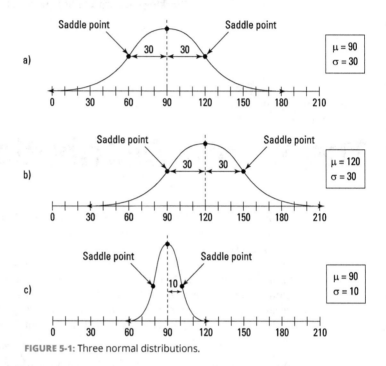

FIGURE 5-1: Three normal distributions.

Note that the *saddle points* (highlighted by arrows in Figure 5-1 on either side of the mean) on each graph are where the graph changes from concave down to concave up. The distance from the mean out to either saddle point is equal to the standard deviation for the normal distribution. For any normal distribution, almost all its values lie within three standard deviations of the mean.

The Standard Normal (Z) Distribution

One very special member of the normal distribution family is called the standard normal distribution, or Z-distribution. The *Z-distribution* is used to help find probabilities and solve other types of problems when working with any normal distribution.

The standard normal (Z) distribution has a mean of zero and a standard deviation of 1; its graph is shown in Figure 5-2. A value on the Z-distribution represents the number of standard deviations the data is above or below the mean; these are called z-scores or z-values. For example, $z = 1$ on the Z-distribution represents a value that is 1 standard deviation above the mean. Similarly, $z = -1$ represents a value that is one standard deviation below the mean (indicated by the minus sign on the z-value).

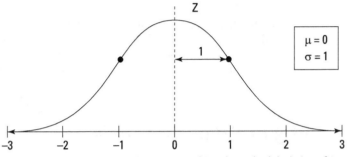

FIGURE 5-2: The Z-distribution has a mean of 0 and standard deviation of 1.

Because probabilities for any normal distribution are nearly impossible to calculate by hand, we use tables to find them. All the basic results you need to find probabilities for any normal distribution can be boiled down into one table based on the standard normal (Z) distribution. This table is called the Z-table and is found in the appendix as Table A-1. All you need is one formula to transform your normal distribution (X) to the standard normal (Z) distribution, and you can use the Z-table to find the probability you need.

The general formula for changing a value of X into a value of Z is $Z = \frac{X - \mu}{\sigma}$. You take your x-value, subtract the mean, and divide by the standard deviation; this gives you its corresponding z-value.

For example, if X is a normal distribution with mean 16 and standard deviation 4, the value 20 on the X-distribution would transform into $20 - 16$ divided by 4, which equals 1. So, the value 20 on the X-distribution corresponds to the value 1 on the Z-distribution. Now use the Z-table to find probabilities for Z, which are equivalent to the corresponding probabilities for X. Table A-1 (appendix) shows the probability that Z is less than any value between -3 and $+3$.

To use the Z-table to find probabilities, do the following:

1. **Go to the row that represents the leading digit of your z-value and the first digit after the decimal point.**

2. **Go to the column that represents the second digit after the decimal point of your z-value.**

3. **Intersect the row and column.**

   That number represents P(Z<z).

For example, suppose you want to look at $P(Z < 2.13)$. Using Table A-1 (appendix), find the row for 2.1 and the column for 0.03. Put 2.1 and 0.03 together as one three-digit number to get 2.13. Intersect that row and column to find the number: 0.9834. You find that $P(Z < 2.13) = 0.9834$.

Finding Probabilities for X

Here are the steps for finding a probability for X:

1. **Draw a picture of the distribution.**

2. **Translate the problem into one of the following: P(X < a), P(X > b), or P(a < X < b). Shade in the area on your picture.**

3. **Transform a (and/or b) into a z-value, using the Z-formula: $Z = \dfrac{X - \mu}{\sigma}$.**

4. **Look up the transformed z-value on the Z-table (see the preceding section) and find its probability.**

5a. **If you have a less-than problem, you're done.**

5b. **If you have a greater-than problem, take one minus the result from Step 4.**

5c. **If you have a between-values problem, do Steps 1–4 for b (the larger of the two values) and then for a (the smaller of the two values), and subtract the results.**

TIP

You need not worry about whether to include an "equal to" in a less-than or greater-than probability because the probability of a continuous random variable equaling one number exactly is zero. (There is no area under the curve at one specific point.)

Suppose, for example, that you enter a fishing contest. The contest takes place in a pond where the fish lengths have a normal distribution with mean $\mu = 16$ inches and standard deviation $\sigma = 4$ inches.

Problem 1: What's the chance of catching a small fish — say, less than 8 inches?

Problem 2: Suppose a prize is offered for any fish over 24 inches. What's the chance of catching a fish at least that size?

Problem 3: What's the chance of catching a fish between 16 and 24 inches?

To solve these problems, first draw a picture of the distribution. Figure 5-3 shows a picture of X's distribution for fish lengths. You can see where each of the fish lengths mentioned in each of the three fish problems falls.

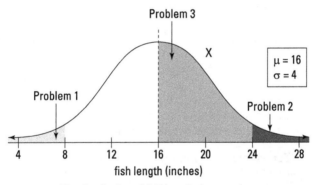

FIGURE 5-3: The distribution of fish lengths in a pond.

Next, translate each problem into probability notation. Problem 1 means find $P(X < 8)$. For Problem 2, you want $P(X < 24)$. And Problem 3 is asking for $P(16 < X < 24)$.

Step 3 says change the x-values to z-values using the Z-formula, $Z = \frac{X - \mu}{\sigma}$. For Problem 1 of the fish example, you have $P(X < 8) = P\left(Z < \frac{8 - 16}{4}\right) = P(Z < -2)$. Similarly for Problem 2, $P(X > 24)$ becomes $P(Z > 2)$. Problem 3 translates from $P(16 < X < 24)$ to $P(0 < Z < 2)$. Figure 5-4 shows a comparison of the X-distribution and Z-distribution for the values $x = 8, 16,$ and 24, which transform into $z = -2, 0,$ and +2, respectively.

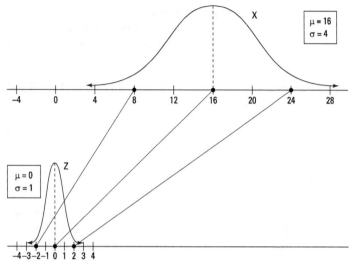

FIGURE 5-4: Transforming numbers on the normal distribution to numbers on the Z-distribution.

Now that you have changed x-values to z-values, you move to Step 4 and find probabilities for those z-values using the Z-table (Table A-1 in the appendix). In Problem 1 of the fish example, you want $P(Z < -2)$; go to the Z-table and look at the row for -2.0 and the column for 0.00, intersect them, and you find 0.0228 — according to Step 5a you're done. So, the chance of a fish being less than 8 inches is equal to 0.0228.

For Problem 2, find $P(Z > 2.00)$. Because it's a "greater-than" problem, this calls for Step 5b. To be able to use the Z-table you need to rewrite this in terms of a "less-than" statement. Because the entire probability for the Z-distribution equals 1, you know $P(Z > 2.00) = 1 - P(Z < 2.00) = 1 - 0.9772 = 0.0228$. So, the chance that a fish is greater than 24 inches is 0.0228. (Note the answers to Problems 1 and 2 are the same because the Z-distribution is symmetric; see Figure 5-3.)

In Problem 3, you find $P(0 < Z < 2.00)$; this requires Step 5c. First find $P(Z < 2.00)$, which is 0.9772 from the Z-table, and then subtract off the part you don't want, which is $P(Z < 0) = 0.500$ from the Z-table. This gives you $0.9772 - 0.500 = 0.4772$. So the chance of a fish being between 16 and 24 inches is 0.4772.

Finding X for a Given Probability

Another type of problem involves finding percentiles for a normal distribution (see Chapter 2 for the rundown on percentiles). That is, you are given the percentage or probability of being below a certain x-value, and you have to find the x-value that corresponds to it. For example, say you want the 50th percentile of the Z-distribution. That is, you want to find the z-value whose probability to its left equals 0.50. Because $P(Z < 0) = 0.5000$ (from Table A-1 of the appendix), you know that 0 is the 50th percentile for Z. But what about other percentiles?

Here are the steps for finding percentiles for a normal distribution X:

1. **If you're given the probability (percent) less than x and you need to find x, you translate this as: Find a where $P(X < a) = p$ (and p is given).** That is, find the pth percentile for X. Go to Step 3.

2. **If you're given the probability (percent) greater than x and you need to find x, you translate this as: Find b where $P(X > b) = p$ (and p is given).** Rewrite this as a percentile (less-than) problem: Find b where $P(X > b) = 1 - p$. This means find the $(1 - p)$th percentile for X.

3. **Find the corresponding percentile for Z by looking in the body of the Z-table (Table A-1 in the appendix) and finding the probability that is closest to p (if you came straight from Step 1) or closest to $1 - p$ (if you came from Step 2).** Find the row and column this number is in (using the table backwards). This is the desired z-value.

4. **Change the z-value back into an x-value (original units) by using $X = \mu + Z\sigma$. (This is the Z-formula, $Z = \dfrac{X - \mu}{\sigma}$, rewritten so X is on the left-hand side.)** You have found the desired percentile for X.

For the fish example, the lengths (X) of fish in a pond have a normal distribution with mean 16 inches and standard deviation 4 inches. Suppose you want to know what length marks the bottom 10 percent of all the fish lengths in the pond. Step 1 says translate the problem; in this case you want to find x such that $P(X < x) = 0.10$. This represents the 10th percentile for X. Figure 5-5 shows a picture of what you need to find in this problem. Now go to Step 3.

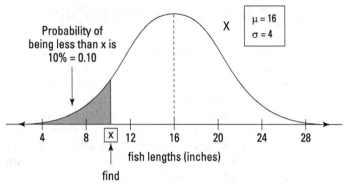

Probability of being less than x is 10% = 0.10

$\mu = 16$
$\sigma = 4$

X

4   8   X   12   16   20   24   28

fish lengths (inches)

find

FIGURE 5-5: Bottom 10 percent of fish in the pond, according to length.

Step 3 says find the 10th percentile for Z. (Although you don't know the x-value that corresponds to a probability of 0.10, you are able find the value of Z that corresponds to 0.10, using the Z-table backwards.) Looking at the Z-table (Table A-1 in the appendix), the probability closest to 0.10 is 0.1003, which falls in the row for $z = -1.2$ and the column for 0.08. The 10th percentile for Z is -1.28. A fish at the bottom 10 percent is 1.28 standard deviation below the mean.

But exactly how long is the fish? In Step 4, you change the z-value back to an x-value (fish length in inches) using the Z-formula solved for X; you get $x = 16 + -1.28 * 4 = 10.88$ inches. So 10.88 inches marks the lowest 10 percent of fish lengths. Ten percent of the fish are shorter than that.

Now suppose you want to find the length that marks the top 25 percent of all the fish in the pond. This means you want to find x where $P(X > x) = 0.25$, so skip Step 1 and go to Step 2. The number you want is in the right tail (upper area) of the X-distribution, with $p = 25$ percent of the probability to the right and $1 - p = 75$ percent to the left. This represents the 75th percentile for X.

TECHNICAL STUFF

Because the Z-table only uses less-than probabilities, you have to rewrite all greater-than probabilities as "one minus" their corresponding less-than probabilities. That is, write everything in terms of percentiles.

Step 3: The 75th percentile of Z is the z-value where $P(Z < z) = 0.75$. Using the Z-table (Table A-1 in the appendix) you find the probability closest to 0.7500 is 0.7486, and its corresponding z-value is in the row for 0.6 and column for 0.07. Put these digits together and get a z-value of 0.67. This is the 75th percentile for Z. In Step 4, change the z-value back to an x-value (length in inches) using the Z-formula solved for X to get $x = 16 + 0.67 * 4 = 18.68$ inches. So, 25 percent of the fish are longer than 18.68 inches (answering the original question). And it's true, 75 percent of the fish are shorter than that.

Normal Approximation to the Binomial

Suppose you flip a fair coin 100 times, and you let X equal the number of heads. What's the probability that X is greater than 60? In Chapter 4, you solve problems like this using the binomial distribution. For binomial problems where n is small, you can either use the direct formula (found in Chapter 4) or the binomial table (Table A-3 in the appendix). However, when n is large, the calculations get unwieldy and the table runs out of numbers. What to do?

Turns out, if n is large enough, you can use the normal distribution to get an approximate answer that's very close to what you would get with the binomial distribution. To determine whether n is large enough to use the normal approximation, two (not just one) conditions must hold:

1. $(n * p) \geq 10$
2. $n * (1 - p) \geq 10$

In general, follow these steps to find the approximate probability for a binomial distribution when n is large:

1. **Verify whether n is large enough to use the normal approximation by checking the two conditions.**

    For the coin-flipping question, the conditions are met since $n * p = 100 * 0.50 = 50$, and $n * (1 - p) = 100 * (1 - 0.50) = 50$, both of which are at least 10. So go ahead with the normal approximation.

2. **Write down what you need to find as a probability statement about X.**

    For the coin-flipping example, find $P(X > 60)$.

3. **Transform the *x*-value to a *z*-value, using the Z-formula,**
$$Z = \frac{X - \mu}{\sigma}.$$

For the mean of the normal distribution, use $\mu = n * p$ (the mean of the binomial), and for the standard deviation σ, use $\sqrt{np(1-p)}$ (the standard deviation of the binomial).

For the coin-flipping example, use $\mu = n * p = 100 * 0.50 = 50$ and $\sigma = \sqrt{np(1-p)} = \sqrt{100 * 0.50(1-0.50)} = 5$.

Now put these values into the Z-formula to get

$Z = \dfrac{60 - 50}{5} = 2$. Now find $P(Z > 2)$.

4. **Proceed as you usually would for any normal distribution. That is, do Steps 4 and 5 described in the earlier section "Finding Probabilities for X."**

For the coin flips, $P(X > 60) = P(Z > 2.00) = 1 - 0.9772 = 0.0228$. The chance of getting more than 60 heads in 100 flips of a coin is about 2.28 percent.

When you use the normal approximation to find a binomial probability, your answer is an approximation (not exact), so be sure you state that. Also show that you checked the necessary conditions for using the normal approximation.

Chapter **6**

Sampling Distributions and the Central Limit Theorem

When you take a sample of data, it's important to realize the results will vary from sample to sample. Statistical results based on samples should include a measure of how much they expect those results to vary from sample to sample. This chapter shows you how to do that by couching everything in terms of the sample means (for numerical data) and applying the same ideas to sample proportions (for categorical data).

Sampling Distributions

Suppose everyone on the planet rolled a single die and recorded the outcome, X. With all those outcomes, we'd have an entire population of values. The graph of these outcomes in the population would represent the distribution of X. Now suppose everyone rolled his or her die 10 times (a sample of size 10) and recorded the average, $\bar{x}$. With all those averages, we'd get an entirely new population — the population of sample means. The graph of this new population would represent the sampling distribution of $\bar{X}$.

When you're talking about a particular sample mean, use the notation $\bar{x}$. When you're talking about the random variable representing any sample mean in general, use the notation $\bar{X}$. A *distribution* is a listing or graph of all possible values of a random variable or a population (such as X) and how often they occur. For example, if you roll a fair die and record the outcome and repeat an infinite number of times, the distribution of X = the outcome, with numbers 1, . . . , 6 appearing with equal frequency. The distribution of X in this case is shown in Figure 6-1a.

Now apply this idea to sample means. Take a sample of values from your random variable X (your population), find the mean of the sample, and repeat over and over again. You now have a new random variable called $\bar{X}$, which takes on a wide range of possible values and has its own distribution.

A listing or graph of all possible values of the sample mean and how often they occur is called the *sampling distribution of the sample mean*. For example if you roll a die 10 times, find the average, and then repeat infinite times, the average will take on values fairly close to 3.5 (halfway between 1 and 6) with values near 3.5 occurring more often than values near 1 or 6. Figure 6-1b shows the actual sampling distribution of $\bar{X}$, the average of 10 rolls of a die.

The term *sampling distribution* is used because data represent averages based on samples, not individual values from a population. As with any other distribution, a sampling distribution has its own shape, center, and measure of variability — the following sections discuss these features.

The mean of a sampling distribution

In the die rolling example, the mean of X (the outcome of a single die) is $\mu_x = 3.5$, as seen in Figure 6-1a. The mean of $\bar{X}$, denoted $\mu_{\bar{x}}$, equals 3.5 as well. The average of a single roll is the same as the average of all possible sample means from 10 rolls.

In general, the mean of this population of all possible sample means is the same as the mean of the entire population. Notationally speaking, you write $\mu_{\bar{x}} = \mu_x$. This makes sense; the average of the averages from all samples is the average of the population that the samples came from.

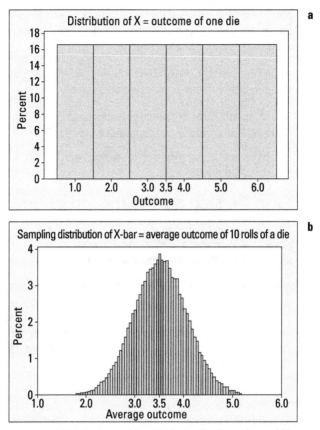

FIGURE 6-1: Distributions of a) individual rolls of one die; and b) average rolls of 10 dice.

TIP

Using subscripts on μ you can distinguish which mean you're talking about. The mean of X (the individuals in the population) or the mean of $\bar{X}$ (all possible sample means from the population) is denoted $\mu_{\bar{x}}$.

Standard error of a sampling distribution

The values in any population deviate from their mean (people have different heights, and so on). Variability in a population of individuals (X) is measured in *standard deviations* (see Chapter 2).

Sample means vary because you're not sampling the whole population, only a subset. Variability in the sample mean ($\bar{X}$) is measured in terms of *standard errors*.

Error here doesn't mean there's been a mistake — it means there is a gap between the population and sample results.

The standard error of the sample means is denoted by $\sigma_{\bar{x}}$. Its formula is $\frac{\sigma_x}{\sqrt{n}}$, where σ_x is population standard deviation and n is sample size. In the next sections, you see the effect each has on the standard error.

Sample size and standard error

Because n is in the denominator of its formula, the standard error decreases as n increases. It makes sense that having more data gives less variation (and more precision) in your results.

A visual can help you see what's happening here with respect to gaining precision in $\bar{X}$ as n increases. Suppose X is the time it takes for a worker to type and send 10 letters of recommendation. Suppose X has a normal distribution with mean 5 minutes and standard deviation 2 minutes. Figure 6-2a shows the picture of the distribution of X.

Now take a random sample of 10 workers, measure their times, and find the average, $\bar{x}$ each time. Repeat this process over and over, and graph all the possible results for all possible samples. Figure 6-2b shows the picture of the distribution of $\bar{X}$. Notice that it's still centered at 10 (which we expected) and that its variability is smaller; the standard error in this case is $\frac{\sigma_x}{\sqrt{n}} = \frac{2}{\sqrt{10}} = 0.63$. The average times are closer to 10 than the individual times shown in Figure 6-2a. That's because average times for 10 individuals don't change as much as individual times do.

Now take random samples of 50 workers and find their means. This sampling distribution is shown in Figure 6-2c. The variation is even smaller here than it was for $n = 10$; the standard error of $\bar{X}$ in this case is $\frac{2}{\sqrt{50}} = 0.28$. The average times here are even closer to 10 than the ones from Figure 6-2b. Larger sample sizes mean more precision and less change from sample to sample.

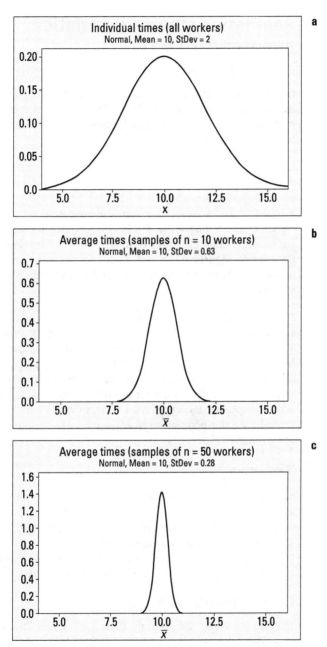

FIGURE 6-2: Distributions of a) individual times; b) average times for 10 individuals; c) average times for 50 individuals.

Population standard deviation and standard error

In the standard error formula for $\bar{X}$, $\frac{\sigma_x}{\sqrt{n}}$ you see that the population standard deviation, σ_x, is in the numerator. That means as the population standard deviation increases, the standard error of the sample means increases. Mathematically this makes sense; how about statistically?

Suppose you have two ponds of fish (call them Pond #1 and Pond #2), and you want to find the average length of all the fish in each pond. Suppose you know that the fish lengths in Pond #1 have a mean of 20 inches and a standard deviation of 2 inches (see Figure 6-3a). Suppose the fish in Pond #2 also average 20 inches, but have a standard deviation of 5 inches (see Figure 6-3b). Comparing Figures 6-3a and 6-3b you see they have the same shape and mean, but the fish in Pond #2 are more variable than in Pond #1.

Now suppose you take a sample of 100 fish from Pond #1, find the mean length of the fish, and repeat this process over and over. Then do the same with Pond #2. Knowing that the fish in Pond #2 have more variability than Pond #1 in the first place, the means of the samples from Pond #2 will have more variability compared to Pond #1 as well. It's harder to estimate the population average when the population varies a lot to begin with — it's much easier to estimate the population average when the population values are similar.

The shape

Now that you know the mean and standard error of $\bar{X}$, the next step is to determine the sampling distribution of $\bar{X}$ (that is, the shape of the distribution of all possible $\bar{X}$'s from all possible samples). There are two cases: 1) the original distribution for X (the population) is normal; and 2) the original distribution for X (the population) is not normal, or is unknown.

Case 1: Distribution of X is normal

If X has a normal distribution, then $\bar{X}$ does too. This is a mathematical statistics result and requires no additional tools to prove. Looking at Figure 6-2, you can see this result is true for the workers' times. Since X is normal, the shape is the same in each graph;

the only thing that changes is the amount concentration around the mean.

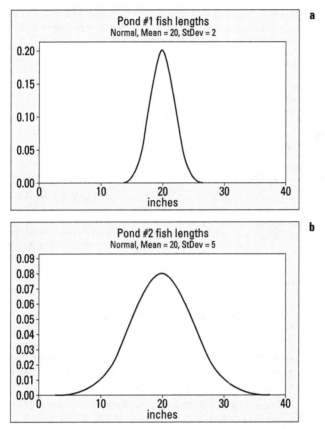

FIGURE 6-3: Distributions of a) fish lengths in Pond #1; b) in Pond #2.

Case 2: Distribution of X is unknown or not normal

If the X distribution is *any* distribution that is not normal, or if its distribution is unknown, you can't automatically say the sample means $(\bar{X})$ have a normal distribution. But you can approximate $\bar{X}$'s distribution with a normal distribution — if the sample size is large enough. This result is due to the *Central Limit Theorem* (CLT). The CLT says that the sampling distribution (shape) of $\bar{X}$ is approximately normal, if the sample size is large enough. And the CLT doesn't care what the distribution of X is!

Formally, for any population with mean μ_x and standard deviation σ_x, the CLT states that:

>> If the distribution of $\bar{X}$ is non-normal or unknown, the sampling distribution of all possible sample means, $\bar{X}$ is *approximately* normal for a sufficiently large sample size.

>> The larger the sample size (n), the closer the distribution of the sample means will be to a normal distribution.

>> Most statisticians agree that if n is at least 30, it will do a reasonable job in most cases.

Two common misconceptions about the CLT:

>> The CLT is only needed when the distribution of X is either non-normal or is unknown. It is not needed if X started out with a normal distribution.

>> The formulas for the mean and standard error of $\bar{X}$ are not due to the CLT. These are just mathematical results that are always true.

Finding Probabilities for $\bar{X}$

After you've established through Case 1 or Case 2 (see previous section) that $\bar{X}$ has a normal or approximately normal distribution, you can find probabilities for $\bar{X}$ by converting the $\bar{x}$-value to a z-value and finding probabilities using the Z-table (Table A-1 in the appendix). The general conversion formula is $Z = \dfrac{\bar{X} - \mu_{\bar{x}}}{\sigma_{\bar{x}}}$. Substituting the appropriate values of the mean and standard error of $\bar{X}$, the conversion formula becomes $Z = \dfrac{\bar{X} - \mu_{\bar{x}}}{\sigma_x \big/ \sqrt{n}}$.

Suppose X is the time it takes a worker to type and send 5 letters of recommendation. Suppose X (the times for all the workers) has a normal distribution and the reported mean is 10 minutes and the standard deviation 2 minutes. You take a random sample of 50 workers and measure their times. What is the chance that their average time is less than 9.5 minutes?

This question translates to finding $P(\bar{X} < 9.5)$. As X has a normal distribution to start with, we know $\bar{X}$ also has a normal distribution. Converting to z-value, we get $= \dfrac{9.5 - 10}{2 / \sqrt{50}} = -1.77$. So we want $P(Z < -1.77)$, which equals 0.0384 from the Z-table (Table A-1 in the appendix). So the chance that these 50 randomly selected workers average less than 9.5 minutes to complete this task is 3.84%.

TIP

Don't forget to divide by the square root of n in the denominator of Z. Always divide by square root of n when the question refers to the *average* of the X- values.

How do you find probabilities for $\bar{X}$ if X is not normal, or is unknown? As a result of the CLT, the distribution of X can be non-normal or even unknown and as long as n is large enough, you can still find approximate probabilities for $\bar{X}$ using the standard normal (Z) distribution and the process described earlier. (That is, convert to a Z-value and find probabilities using the Z-table, Table A-1, appendix.)

TECHNICAL
STUFF

When you do have to use the CLT to find a probability for $\bar{X}$, you need to say that your answer is an approximation and that you've got a large enough n to proceed because of the CLT. (If n is not large enough for the CLT, you use the t-distribution in many cases — see Chapter 9.)

The Sampling Distribution of the Sample Proportion

The Central Limit Theorem (CLT) doesn't apply only to sample means. You can also use it with other statistics, including sample proportions. The *population proportion, p,* is the proportion of individuals in the population that have a certain characteristic of interest based on a binomial random variable (see Chapter 4). The *sample proportion,* denoted $\hat{p}$, is the proportion of individuals in the sample that have that same characteristic of interest. The sample proportion is the number of individuals in the sample who have that characteristic of interest divided by the total sample size (n). If you take a sample of 100 students and find

60 freshman, the sample proportion for freshman is $60/100 = 0.60$. This section examines the sampling distribution of all possible sample proportions, $\hat{p}$, from samples of size n from a population.

The sampling distribution of $\hat{p}$ has these properties:

>> Its mean is the population proportion, denoted by p.

>> Its standard error is $\sqrt{\dfrac{p(1-p)}{n}}$. (Note that because n is in the denominator, standard error decreases as n increases.)

>> Its shape is *approximately* normal, provided that the sample size is large enough. This is due to the CLT. That means you can use the normal distribution to find probabilities for $\hat{p}$. (See Chapter 5 for more.)

>> The larger the sample size (n), the closer the distribution of sample proportions is to a normal distribution.

TECHNICAL STUFF

How large is large enough for the CLT to work for categorical data? Most statisticians agree that both np and $n(1-p)$ should be greater than or equal to 10. You want the average number of successes (np) and the average number of failures $n(1-p)$ to be at least 10. (Note the second condition involves $n(1-p)$, not $np(1-p)$, the variance of the binomial distribution.)

What proportion of students need math help?

Suppose you want to know what proportion of incoming college students would like help in math. A student survey accompanies the ACT test each year, and one of the questions is whether the student would like some help with math skills. Assume (through past research) that 38% of the students taking the ACT respond yes. That means $p = 0.38$ in this case.

The original data have a binomial distribution where success = would like help. The yes responses (p) and no responses ($1-p$) for the population are shown in Figure 6-4 as a bar graph. (See Chapter 3 for more on bar graphs.)

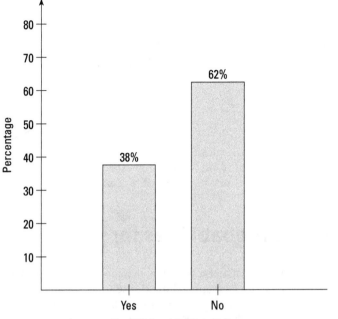

FIGURE 6-4: Population percentages for responses to ACT math-help question.

Now take all possible samples of size 1,000 from this population and find the proportion in each who said they needed math help. The distribution of these sample proportions is in Figure 6-5. It has an approximate normal distribution with mean $p = 0.38$ and standard error equal to

$$\sqrt{\frac{p(1-p)}{n}} = \sqrt{\frac{0.38(1-0.38)}{1,000}} = 0.015$$

(or about 1.5%). This approximation is valid because the two conditions for the CLT are met: 1) $np = 1,000(0.38) = 380$ (which is at least 10); and 2) $n(1-p) = 1,000(0.62) = 620$ (also at least 10).

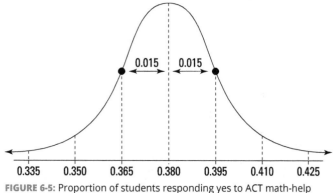

FIGURE 6-5: Proportion of students responding yes to ACT math-help question for samples of size 1,000.

Finding Probabilities for $\hat{p}$

For the ACT test example, suppose it's reported that 0.38 or 38% of all the students taking the ACT test would like math help. Suppose you took a random sample of 1,000 students. What is the chance that more than 40 percent of them say they need help?

What the question wants is the probability that the sample proportion, $\hat{p}$ is greater than 0.40; that is, $P(\hat{p} > 0.40)$. This question is answered using the normal approximation for $\hat{p}$ described in the previous section, given the stated conditions are met.

You first check the conditions: 1) is np at least 10? Yes because $1,000 * 0.38 = 380 = 38$; 2) is $n(1-p)$ at least 10? Again yes because $1.000 * (1-0.38) = 620$ checks out. So you can use the normal approximation to answer the question.

You make the conversion of the $\hat{p}$-value to a z-value using $Z = \dfrac{\hat{p}-p}{\sqrt{\dfrac{p(1-p)}{n}}}$ to get $Z = \dfrac{0.40-0.38}{\sqrt{\dfrac{0.38(1-0.38)}{1,000}}} = 1.30$. Now you find $P(Z > 1.38) = 1 - 0.9032 = 0.0968$. So if 38 percent of students wanted help, the chance of taking a sample of 1,000 students and getting more than 40 percent needing help is approximately 0.0968 (by the CLT).

Comparing sample results to a claim about the population is called *hypothesis testing*. Because the chance of getting more than 40% of the students in our sample who requested help is 0.0968, we wouldn't reject the claim that 38% of the population of all ACT takers request help. To reject this claim, most statisticians would want this probability to be less than 0.05 (see Chapter 8 for more on hypothesis testing).

Chapter **7**

Confidence Intervals

I n this chapter, you find out how to build, calculate, and interpret confidence intervals, and you work through the formulas involving one or two population means or proportions. You also get the lowdown on some of the finer points of confidence intervals: what makes them narrow or wide, what makes you more or less confident in their results, and what they do and don't measure.

Making Your Best Guesstimate

A *confidence interval* (abbreviated CI) is used for the purpose of estimating a population parameter (a single number that describes a population) by using statistics (numbers that describe a sample of data). For example, you might estimate the average household income (parameter) based on the average household income from a random sample of 1,000 homes (statistic). However, because sample results will vary (see Chapter 6), you need to add a measure of that variability to your estimate. This measure of variability is called the margin of error, the heart of a confidence interval. Your sample statistic, plus or minus your margin of error, gives you a range of likely values for the parameter — in other words, a confidence interval.

The margin of error is the amount of "plus or minus" that is attached to your sample result when you move from discussing the sample itself to discussing the whole population that it represents; that's why the general formula for the margin of error contains a "±" in front of it.

For example, say the percentage of kids who like baseball is 40 percent, plus or minus 3.5 percent. That means the percentage of kids who like baseball is somewhere between $40\% - 40\% - 3.5\% = 36.5\%$ and $40\% + 3.5\% = 43.5\%$. The lower end of the interval is your statistic minus the margin of error, and the upper end is your statistic plus the margin of error.

TECHNICAL STUFF

The margin of error is not the chance a mistake was made; it measures variation in the random samples due to chance. Because you didn't get to sample everybody in the population, you expect your sample results to be "off" by a certain amount, just by chance. You acknowledge that your results could change with subsequent samples, and that they're only accurate to within a certain range, which is the margin of error.

To estimate a parameter with a confidence interval:

1. **Choose your confidence level and your sample size (see details later in this chapter).**

2. **Select a random sample of individuals from the population.**

3. **Collect reliable and relevant data from the individuals in the sample.**

   See Chapter 12 for survey data and Chapter 13 for data from experiments.

4. **Summarize the data into a statistic (for example, a sample mean or proportion).**

5. **Calculate the margin of error.** (Details later in this chapter.)

6. **Take the statistic plus or minus the margin of error to get your final estimate of the parameter.**

   This is called a *confidence interval* for that parameter.

For example, the formula for a confidence interval for the mean of a population is $\bar{x} \pm z * \frac{\sigma}{\sqrt{n}}$; the statistic here is $\bar{x}$ (the sample mean), and the margin of error is the piece following the plus/minus sign: $\pm z * \frac{\sigma}{\sqrt{n}}$. (This formula is fully broken down in the section "Confidence Interval for a Population Mean.")

The Goal: Small Margin of Error

The ultimate goal when making an estimate using a confidence interval is to have a small margin of error. The narrower the interval, the more precise the results are.

For example, suppose you're trying to estimate the percentage of semitrucks on the interstate between the hours of 12 a.m. and 6 a.m., and you come up with a 95% confidence interval that claims the percentage of semis is 50%, plus or minus 40%. Wow, that narrows it down! (Not.) You've defeated the purpose of trying to come up with a good estimate — the confidence interval is much too wide. You'd rather say something like: a 95% confidence interval for the percentage of semis on the interstate between 12 a.m. and 6 a.m. is 50%, plus or minus 3% (thus between 47% and 53%).

How do you go about ensuring that your confidence interval will be narrow enough? You certainly want to think about this issue before collecting your data; after the data are collected, the width of the confidence interval is set.

Three factors affect the size of the margin of error:

>> The confidence level

>> The sample size

>> The amount of variability in the population

These three factors all play important roles in influencing the width of a confidence interval. In the following sections, you see how.

TECHNICAL STUFF

Note that the sample statistic itself (for example, 50% of vehicles in the sample are semis) isn't related to the width of the confidence interval. The statistic only determines the midpoint of the confidence interval, not its width.

Choosing a Confidence Level

Variability in sample statistics is measured in standard errors. A *standard error* is very similar to the standard deviation of a data set or a population. The difference is that a standard error measures the variation among all the possible values of the statistic (for example, all the possible sample means) while a standard deviation of a population measures the variation among all possible values within the population itself. (See Chapter 6 for all the information on standard errors.)

The *confidence level* of a confidence interval corresponds to the percentage of the time your result would be correct if you took numerous random samples. Typical confidence levels are 95% or 99% (many others are also used). The confidence level determines the number of standard errors you add and subtract to get the percentage confidence you want.

When working with means and proportions, if the proper conditions are met, the number of standard errors to be added and subtracted for a given confidence level is based on the standard normal (Z-) distribution, and is labeled $z*$. The higher the confidence level, the more standard errors need to be added and subtracted, hence a higher $z*$-value. For 95% confidence, the $z*$-value is 1.96, and for 99% confidence, $z*$-value is 2.58. Some of the more commonly used confidence levels, along with their corresponding $z*$-values, are given in Table 7-1.

TABLE 7-1 $z*$-values for Selected (Percentage) Confidence Levels

Percentage Confidence	$z*$-value
80	1.28
90	1.64
95	1.96
98	2.33
99	2.58

Using stat notation, you can write a confidence level as $(1-\alpha)$, where α represents the percentage of confidence intervals that are incorrect (don't contain the population parameter by random chance). So if you want a 95 percent confidence interval, $\alpha = 0.05$. This number α is also related to the chance of making a Type I error in a hypothesis test (see Chapter 8).

Factoring in the Sample Size

The relationship between margin of error and sample size is simple: As the sample size increases, the margin of error decreases. This confirms what you hope is true: The more information you have, the more accurate your results are going to be. (That, of course, assumes that it's good, credible information — see Chapters 12 and 13.)

Looking at the formula for standard error for the sample mean, $\frac{\sigma}{\sqrt{n}}$ (from Chapter 6) notice that it has an n in the denominator of a fraction; this is the case for most any standard error formula. As n increases, the denominator of this fraction increases, which makes the overall fraction get smaller. That makes the margin of error, $z * \frac{\sigma}{\sqrt{n}}$, smaller and results in a narrower confidence interval.

Here's where a large sample size really comes in handy. When you need a high level of confidence, you have to increase the $z *$-value and, hence, the margin of error. This makes your confidence interval wider (not good). But you can offset this wider confidence interval by increasing the sample size and bringing the margin of error back down, thus narrowing the confidence interval. The increase in sample size allows you to still have the confidence level you want, but also ensures that the width of your confidence interval will be small (which is what you ultimately want).

You can determine the sample size you need to achieve a certain margin of error before you start a study. When estimating a population mean, you can use the following sample size formula: $n\left\lceil \left(\frac{z * \sigma}{\text{MOE}}\right)^2 \right\rceil$, where *MOE* is your desired margin of error; σ is the population standard deviation; and $z*$ is the value on the Z-distribution that corresponds to the confidence level you want (Table 7-1).

Notice that the bracket notation on the outside of the equation for *n* has a flat ledge on top and no ledge on the bottom. That means you are supposed to round up your result to the "next greatest integer." In other words, always round up your answer to the next integer if you have anything after the decimal point — even 107.01 is rounded up to 108. This ensures that you won't exceed the margin of error you need.

If the population standard deviation, σ, is unknown, you can do a *pilot* study (a small study before the full blown study) and use its sample standard deviation (*s*) as a substitute for σ. At that point you would use the appropriate value on the *t*-distribution with $n-1$ degrees of freedom, rather than z^*. (See Chapter 9 for info on the *t*-distribution.)

When your statistic is a sample proportion or percentage (such as the proportion of females, or the percentage of semis) a quick-and-dirty way to figure margin of error is to take 1 divided by the square root of *n* (the sample size). Try different values of *n* and see how the margin of error is affected.

Approximately what sample size is needed to have a narrow confidence interval with respect to polls? Using the formula in the preceding paragraph, you can make some quick comparisons. A survey of 100 people will have a margin of error of about $\frac{1}{\sqrt{100}} = 0.10$, or plus or minus 10% (which is fairly large). However, if you survey 1,000 people, your margin of error decreases dramatically, to plus or minus $\frac{1}{\sqrt{1,000}}$, or about 3%. A survey of 2,500 people in the U.S. results in a margin of error of plus or minus 2%. This sample size gives amazing accuracy when you think about how large the U.S. population is (well over 300 million).

Keep in mind, however, that you don't want to go too high with your sample size because there is a point where you start having a diminished return. For example, moving from a sample size of 2,500 to 5,000 narrows the margin of error of the confidence interval to about 1.4%, down from 2%. Each time you survey one more person, the cost of your survey in terms of money and time increases, so adding another 2,500 people to the survey just to narrow the interval by less than six-tenths of 1% may not be worthwhile.

Real accuracy depends on the quality of the data as well as on the sample size. A large sample size that has a great deal of bias (see Chapter 12) may appear to have a narrow confidence interval but actually means nothing. It's better to have a smaller sample size that contains good data than a larger sample size with a lot of bias.

Counting on Population Variability

Another factor influencing variability in sample results is the variability (standard deviation) within the population itself. For example, in a population of houses in a large city like Columbus, Ohio, you see a large amount of variability in price. This variability in house price over the whole city will be higher than the variability in house price if your population was limited to a certain housing development in Columbus (where the houses are likely to be similar to each other).

As a result, if you take a sample of houses from the entire city of Columbus and find the average price, the margin of error will be larger than if you take a sample from one single housing development in Columbus. So you'll need to sample more houses from the entire city of Columbus in order to have the same amount of accuracy that you would get from a single housing development.

You can also look at it mathematically. Variability is measured in terms of standard errors/deviations. Notice that the population standard deviation, σ, appears in the numerator of the standard error of the sample mean, $\frac{\sigma}{\sqrt{n}}$. As σ (numerator) increases, the standard error (entire fraction) increases. A larger standard error means a larger margin of error and a wider confidence interval.

More variability in the original population increases the margin of error, making the confidence interval wider. However, don't let that discourage you. This increase can be offset by increasing the sample size. (Remember the sample size, n, appears in the denominator of the standard error formula, $\frac{\sigma}{\sqrt{n}}$, so an increase in n results in a decrease in the margin of error.)

Confidence Interval for a Population Mean

When the characteristic that's being measured (such as income, IQ, price, height, quantity, or weight) is *numerical,* people often want to estimate the *mean* (average) value for the population. You estimate the population mean by using a sample mean plus or minus a margin of error. The result is a *confidence interval for a population mean μ.*

The formula for a CI for a population mean is

$$\bar{x} \pm z^* \frac{\sigma}{\sqrt{n}}$$

where $\bar{x}$ is the sample mean; σ is the population standard deviation; n is the sample size; and z^* is the appropriate value from the Z-distribution for your desired confidence level (see Table 7-1 for values of z^* for given confidence levels).

For example, suppose you work for the Department of Natural Resources and you want to estimate, with 95% confidence, the mean (average) length of the walleyes in a fish hatchery pond. (Assume the population standard deviation (σ) is 2.3 inches.) Because you want a 95% confidence interval, your z^*-value is 1.96. Suppose you take a random sample of $n = 100$ walleyes and find the average length ($\bar{x}$) is 7.5 inches. To find the margin of error, multiply 1.96 times 2.3 divided by the square root of 100 to get plus or minus $1.96*(2.3/10) = 0.45$ inches.

Your 95% confidence interval for the mean length of the walleyes in this fish hatchery pond is 7.5 inches plus or minus 0.45 inches. (The lower end of the interval is $7.5 - 0.45 = 7.05$ inches; the upper end is $7.5 + 0.45 = 7.95$ inches.) You can say that a range of likely values for the average length of the walleyes in this entire pond is between 7.05 and 7.95 inches, based on your sample, with a confidence level of 95%.

When your sample size is small (under 30), you use the appropriate value on the t-distribution with $n-1$ degrees of freedom instead of z^* (see Table A-2 in the appendix).

You can also use a confidence interval for one population mean to analyze the average difference in paired data from one population. For example, suppose you want to estimate the average effect of a certain drug on blood pressure. You take one sample of patients, measure their blood pressure before and after taking the drug, and record the differences in blood pressure. (This type of experiment is called a *matched-pairs design*; see Chapter 13.) These differences represent a single sample from a single population, so a confidence interval for one population mean can be used to estimate the average difference in blood pressure due to the drug.

Confidence Interval for a Population Proportion

When a characteristic being measured is categorical — for example, opinion on an issue (support, oppose, or neutral), or type of behavior (do/don't wear a seatbelt while driving), people often want to estimate the proportion (or percentage) of people in the population who fall into a certain category of interest. Examples include the percentage of people in favor of a four-day workweek, or the proportion of drivers who don't wear seatbelts. In each of these cases, the object is to estimate a population proportion using a sample proportion plus or minus a margin of error. The result is called a *confidence interval for a population proportion, p.*

The formula for a CI for a population proportion, *p*, is

$$\hat{p} \pm z^* \sqrt{\frac{\hat{p}(1-\hat{p})}{n}}$$

where $\hat{p}$ is the sample proportion; *n* is the sample size; and z^* is the appropriate value from the standard normal (Z-) distribution for your desired confidence level. (Note that a *sample proportion* is the proportion of individuals in the sample that had the characteristic of interest.)

For example, suppose you want to estimate the percentage of the time you get a red light at a certain intersection. If you want a 95% confidence interval, your z^*-value is 1.96. You take a random sample of 100 different trips through this intersection, and you find that you hit a red light 53 times, so $\hat{p} = 53/100 = 0.53$. Take 0.53 times $(1-0.53)$ and divide by 100 to get $0.249/100 = 0.00249$.

Take the square root to get 0.0499 or 0.05. The margin of error is, therefore, plus or minus $1.96 * 0.05 = 0.098$.

Your 95% confidence interval for the percentage of times you will ever hit a red light at that particular intersection is 0.53 (or 53%) plus or minus 0.098. The lower end of the interval is $0.530 - 0.098 = 0.43$ or 43%; the upper end is $0.530 + 0.098 = 0.63$ or 63%. You conclude the overall percentage of the times you should expect to hit a red light at this intersection is somewhere between 43% and 63%, based on your sample, with a confidence level of 95%.

Confidence Interval for the Difference of Two Means

The goal of many surveys and medical studies is to compare two populations, such as males versus females or Republicans versus Democrats. When the characteristic being compared is numerical (for example, height, weight, or income), the object of interest is the amount of difference in the means (averages) for the two populations. For example, you may want to compare the difference in average age of Republicans versus Democrats, or the difference in average incomes of men versus women. You estimate the difference between two population means by taking a sample from each population and using the difference of the two sample means, plus or minus a margin of error. The result is a *confidence interval for the difference of two population means, $\mu_1 - \mu_2$.*

The formula for a CI for the difference between two population means is

$$\bar{x} - \bar{y} \pm z * \sqrt{\frac{\sigma_1^2}{n_1} + \frac{\sigma_2^2}{n_2}}$$

where $\bar{x}$ and $\bar{y}$ are the sample means, respectively; n_1 and n_2 are the sample sizes; σ_1 and σ_2 are the population standard deviations; and $z*$ is the appropriate value from the standard normal (Z-) distribution for your desired confidence level (see Table 7-1 for values of $z*$ for certain confidence levels).

If one or both of the sample sizes are small (less than 30), you use the appropriate value on the t-distribution with $n_1 + n_2 - 2$ degrees of freedom instead of z^* (see Table A-2 in the appendix).

Suppose you want to estimate with 95% confidence the difference between the mean (average) lengths of cobs from two varieties of sweet corn (allowing them to grow the same number of days under the same conditions). Call the two varieties Corn-e-stats and Stats-o-sweet.

Suppose your random sample of 100 cobs of the Corn-e-stats variety averages 8.5 inches, with a standard deviation of 2.3 inches, and your random sample of 110 cobs of Stats-o-sweet averages 7.5 inches, with a standard deviation of 2.8 inches. That is, $\bar{x} = 8.5, s_1 = 2.3$, and $n_1 = 100$ from the Corn-e-stats; and $\bar{y} = 7.5$, $s_2 = 2.8$, and $n_2 = 110$ from the Stats-o-sweet.

Notice the population standard deviations are unknown; when this is the case you substitute the appropriate value from the t-distribution with $n_1 + n_2 - 2$ degrees of freedom for z^*. In this case, the degrees of freedom are $100 + 110 - 2 = 208$; with this many degrees of freedom, the t- and Z-distributions are approximately equal (see Chapter 9), and you use 1.96 for the appropriate value of t anyway (see last row of Table A-2 in the appendix).

The difference between the sample means $\bar{x} - \bar{y}$ is $8.5 - 7.5 = +1$ inch. The average for Corn-e-stats minus the average for Stats-o-sweet is positive, making Corn-e-stats the larger of the two varieties, in terms of this sample. Is that difference enough to generalize to the entire population, though? That's what this confidence interval is going to help you decide.

To calculate the margin of error, square s_1 (2.3) to get 5.29 and divide by 100 to get 0.0529; then square s_2 (2.8) and divide by 110 to get $7.84 / 110 = 0.0713$. The sum is $0.0529 + 0.0713 = 0.1242$; the square root is 0.3524. Multiply 1.96 times 0.3524 to get 0.69 inch, the margin of error.

Your 95% confidence interval for the difference between the average lengths for these two varieties of sweet corn is 1 inch, plus or minus 0.69 inch. (The lower end of the interval is $1 - 0.69 = 0.31$ inch; the upper end is $1 + 0.69 = 1.69$ inches.) You conclude that the cobs of the Corn-e-stats variety are longer, on average, than the Stats-o-sweet variety, by between 0.31 and 1.69 inches, with a 95% level of confidence.

TECHNICAL STUFF

Notice all the values in this interval are positive. That's why you conclude one brand is longer than the other (according to your data). If some of the values in the confidence interval were positive and some were negative, you wouldn't conclude one was longer than the other on average.

Also note that there is a difference between the "difference in the means" and the "mean of the differences." If you're looking at pairs of data (such as pre-test versus post-test) and are examining the differences, you only have one data set and one population. Use the methods in the "Confidence Interval for a Population Mean" section to find a confidence interval for the "mean difference." If you're examining the difference in the means of two separate populations (such as males versus females), use the methods in this section to find a confidence interval for the "difference of two means."

TIP

Notice that you could get a negative value for $\bar{x} - \bar{y}$. For example, if you had switched the two varieties of corn, you would have gotten -1 for this difference. That's fine; just remember which group is which. A positive difference means the first group has a larger value than the second group; a negative difference means the first group has a smaller value than the second group. If you want to avoid negative values, always make the group with the larger value your first group — all your differences will be positive.

Confidence Interval for the Difference of Two Proportions

When two populations are compared regarding some categorical variable (such as comparing males to females regarding their opinion of a four-day workweek), you estimate the difference between the two population proportions. You do this by taking the difference in their corresponding sample proportions (one from each population) plus or minus a margin of error. The result is called a *confidence interval for the difference of two population proportions, $p_1 - p_2$.*

The formula for a confidence interval for the difference between two population proportions is:

$$\left(\hat{p}_1 - \hat{p}_2\right) \pm z^* \sqrt{\frac{\hat{p}_1\left(1 - \hat{p}_1\right)}{n_1} + \hat{p}_2 \frac{\left(1 - \hat{p}_2\right)}{n_2}}$$

where $\hat{p}_1$ and n_1 are the sample proportion and sample size of the first sample; $\hat{p}_2$ and n_2 are the sample proportion and sample size of the second sample; and $z*$ is the appropriate value from the standard normal (Z-) distribution for your desired confidence level (see Table 7-1 for $z*$-values).

Suppose you work for the Las Vegas Chamber of Commerce and you want to estimate with 95% confidence the difference between the proportion of females versus males who have ever gone to see an Elvis impersonator. Suppose your random sample of 100 females includes 53 females who have seen an Elvis impersonator, so $\hat{p}_1$ is $53/100 = 0.53$; and your random sample of 110 males includes 37 males who have ever seen an Elvis impersonator, so $\hat{p}_2$ is $37/100 = 0.34$. Because you want a 95% confidence interval, your $z*$-value is 1.96. Using the formula for the confidence interval for the difference of two proportions, you get the following:

$$(0.53 - 0.34) \pm 1.96 \sqrt{\frac{0.53(1-0.53)}{100} + \frac{0.34(1-0.34)}{110}} =$$
$$0.19 \pm 1.96 \sqrt{0.003 + 0.002}$$

which equals 0.19 plus or minus 0.13.

TIP

While performing any calculations involving sample percentages, you must use the decimal form. After the calculations are finished, you may convert to percentages by multiplying by 100.

Your 95% confidence interval for the difference between the percentage of females who have seen an Elvis impersonator and the percentage of males who have seen an Elvis impersonator is 19%, plus or minus 13%. The lower end of the interval is $0.19 - 0.13 = 0.06$ or 6%; the upper end is $0.19 + 0.13 = 0.32\%$. You conclude that a higher percentage of females have seen an Elvis impersonator (compared to males), and the difference is somewhere between 6% and 32%, with a 95% level of confidence. (Note this interval is quite wide; if you increase the sample sizes, the margin of error will decrease because n_1 and n_2 are in the denominator of the formula for the margin of error.)

Interpreting Confidence Intervals

The big idea of a confidence interval is that it presents a range of likely values for the population parameter, based on one random sample, with a certain confidence level (such as 95%).

This sounds fairly straightforward, but there are some intricacies that can lead to incorrect interpretation of the results. This section helps untangle the confusion that can occur when interpreting a confidence interval.

Consider a survey conducted by the Gallup Organization (a world leader in the survey business). Suppose it samples 1,000 people at random from the United States, and the results show that 520 people (52%) think the president is doing a good job. Gallup reports this survey has a margin of error of plus or minus 3%. So far, you know that a majority of the 1,000 people in this sample approve of the president, but can you say this opinion carries over to a majority of *all* Americans?

If 52% of those sampled approve of the president, you can expect the percentage of all Americans who approve of the president to be 52%, plus or minus 3.0%. That is, a range of likely values is between $52\% - 3\% = 49\%$ and $52\% + 3\% = 55\%$. To report the results from this poll, you would say, "Based on my sample, 52% of all Americans approve of the president, plus or minus a margin of error of 3.0 percent, with a confidence level of 95%."

How does a polling organization report its results? Here's how Gallup does it:

> *"Based on the total sample of adults in (this) survey, we are 95% confident that the margin of error for our sampling procedure and its results is no more than ± 3.0 percentage points."*

TECHNICAL STUFF

Notice that 49% (the lower end of the range of likely values) is less than 50%. So you really can't say that a majority of the American people support the president, based on this sample. You can only say that between 49% and 55% of all Americans support the president.

Now comes the subtle but very important point regarding how to interpret a confidence interval. When one particular confidence interval is calculated, do not include a probability statement about your particular result when you draw your conclusions. That is, it's wrong to say "I am 95% confident that the population mean is between XXX and XXX." Once your sample has been selected and your confidence interval is calculated, it either contains the population parameter or it doesn't; there is no probability involved.

Bottom line: The confidence level (in this case, 95%) does not apply to a single confidence interval.

So how do you interpret the 95%? It goes back to the definition of a confidence level. A *confidence level* is the percentage of all possible samples of size *n* whose confidence intervals contain the population parameter. When taking many random samples from a population, you know that some samples (in this case, 95% of them) will represent the population, and some won't (in this case, 5% of them) just by random chance. Random samples that represent the population will result in confidence intervals that contain the population parameter (that is, they are correct); and those that do not represent the population will result in confidence intervals that are not correct.

For example, if you randomly sample 100 exam scores from a large population, you might get more low scores than you should in your sample just by chance, and your confidence interval will be too low; or you might get more high scores than you should in your sample just by chance, and your confidence interval will be too high. These two confidence intervals won't contain the population parameter, but with a 95% confidence level, this type of error (called *sampling error*) should only happen 5% of the time.

TECHNICAL STUFF

Confidence level (such as 95%) represents the percentage of all possible random samples of size *n* that typify the population and hence result in correct confidence intervals. It isn't the probability of a single confidence interval being correct.

Another way of thinking about the confidence level is to say that if the organization took a sample of 1,000 people over and over again and made a confidence interval from its results each time, 95 percent of those confidence intervals would be right. (You just have to hope that yours is one of those right results.)

To correctly interpret your particular confidence interval you can say "A range of likely values for the population mean is XXX to XXX, with a confidence level of 95%." Or you could say it like the Gallup Organization does:

"For these results, one can say with 95% confidence that the maximum amount of sampling (margin of) error is plus or minus XXX."

It's all about the sampling process, not a single sample.

Spotting Misleading Confidence Intervals

There are two possible reasons that a confidence interval is incorrect (does not contain the population parameter). First, it can be incorrect by random chance because the random sample it came from didn't represent the population; or second, it can be incorrect because the data that went into it weren't any good. I discuss the first situation in the previous section, and it can't be prevented. The second situation can be prevented (or at least minimized) through good data-collection practices.

A good slogan to remember when examining statistical results is *"garbage in = garbage out."* No matter how nice and scientific someone's confidence interval may look, the formula that was used to calculate it doesn't have any idea of the quality of the data that went into it. It's up to you to check it out. For example, if the data for the confidence interval was based on a *biased* sample (one that favored certain people over others); a bad design; bad data-collection procedures; or misleading questions, the margin of error is suspect — if the bias is bad enough, the results will be bogus.

For example, suppose a total of 50,000 people were surveyed on a certain issue. This incredibly high sample size sounds great — until you realize they were all visitors to a certain website. The tiny reported margin of error is a result of the huge n, yet it means nothing because it is based on biased data that didn't come from a random sample. Of course, some people will go ahead and report it anyway, so you're left to determine whether the results are based on good information or garbage. If garbage, you know what to do about the margin of error: Ignore it.

Before I get on too high of a horse here, it's important to note that even the best of surveys can still contain a little bias. The Gallup Organization addresses the issue of what margin of error does and does not measure in the follow disclaimer added to its reports:

> *"In addition to sampling error, question wording and practical difficulties in conducting surveys can introduce error or bias into the findings of public opinion polls."*

What Gallup is saying is that besides the error that happens in random samples just by chance, surveys can have additional errors or bias due to things like missing data from people who don't respond, or phone numbers no longer in service. Margin of error cannot measure the extent of those types of nonsampling errors. However, a good survey design like Gallup can go a long way toward helping minimize bias and get credible results. (See Chapter 12 for full details on doing good surveys.)

Chapter **8**
Hypothesis Tests

ypothesis testing is a statistician's way of trying to confirm or deny a claim about a population using data from a sample. For example, you might read on the Internet that the average price of a home in your city is $150,000 and wonder if that number is true for the whole city. Or you hear that 65% of all Americans are in favor of a smoking ban in public places — is this a credible result? In this chapter, I give you the big picture of hypothesis testing as well the details for hypothesis tests for one or two means or proportions. And I examine possible errors that can occur in the process.

Doing a Hypothesis Test

A *hypothesis test* is a statistical procedure that's designed to test a claim. Typically, the claim is being made about a population parameter (one number that characterizes the entire population). Because parameters tend to be unknown quantities, everyone wants to make claims about what their values may be. For example, the claim that 25% (or 0.25) of all women have varicose veins is a claim about the proportion (that's the *parameter*) of all women (that's the *population*) who have varicose veins.

Identifying what you're testing

To get more specific, the varicose vein claim is that the parameter, the population proportion (p), is equal to 0.25. (This claim is called the *null hypothesis*.) If you're out to test this claim, you're questioning the claim and have a hypothesis of your own (called the *research hypothesis,* or *alternative hypothesis*). You may hypothesize, for example, that the actual proportion of women who have varicose veins is lower than 0.25, based on your observations. Or, you may hypothesize that due to the popularity of high-heeled shoes, the proportion may be higher than 0.25. Or, if you're simply questioning whether the actual proportion is 0.25, your alternative hypothesis is, "No, it isn't 0.25."

In addition to testing hypotheses about categorical variables (having or not having varicose veins is a categorical variable), you can also test hypotheses about numerical variables, such as the average commuting time for people working in Los Angeles or their average household income. In these cases, the parameter of interest is the population average or mean (denoted μ). Again, the claim is that this parameter is equal to a certain value, versus some alternative.

Setting up the hypotheses

Every hypothesis test contains two hypotheses. The first hypothesis is called the *null hypothesis,* denoted H_o. The null hypothesis always states that the population parameter is *equal* to the claimed value. For example, if the claim is that the average time to make a name-brand ready-mix pie is 5 minutes, the statistical shorthand notation for the null hypothesis in this case would be as follows: $H_o : \mu = 5$.

What's the alternative?

Before actually conducting a hypothesis test, you have to put two possible hypotheses on the table — the null hypothesis is one of them. But, if the null hypothesis is found not to be true, what's your alternative going to be? Actually, three possibilities exist for the second (or alternative) hypothesis, denoted H_a. Here they are, along with their shorthand notations in the context of the example:

>> The population parameter is *not equal* to the claimed value ($H_a : \mu \neq 5$).

>> The population parameter is *greater than* the claimed value ($H_a : \mu > 5$).

>> The population parameter is *less than* the claimed value ($H_a : \mu < 5$).

Which alternative hypothesis you choose in setting up your hypothesis test depends on what you're interested in concluding, should you have enough evidence to refute the null hypothesis (the claim). For example, if you want to test whether or not a company is correct in claiming its pie takes 5 minutes to make, you use the not-equal-to alternative. Your hypotheses for that test would be $H_o : \mu = 5$ versus $H_o : \mu \neq 5$.

If you only want to see whether the time turns out to be greater than what the company claims (that is, the company is falsely advertising its prep time), you use the greater-than alternative, and your two hypotheses are $H_o : \mu = 5$ versus $H_a : \mu > 5$. Suppose you work for the company marketing the pie, and you think the pie can be made in less than 5 minutes (and could be marketed by the company as such). The less-than alternative is the one you want, and your two hypotheses would be $H_o : \mu = 5$ versus $H_a : \mu < 5$.

Knowing which hypothesis is which

How do you know which hypothesis to put in H_o and which one to put in H_a? Typically, the null hypothesis says that nothing new is happening; the previous result is the same now as it was before, or the groups have the same average (their difference is equal to zero). In general, you assume that people's claims are true until proven otherwise.

TIP

Hypothesis tests are similar to jury trials, in a sense. In a jury trial, H_o is similar to the not-guilty verdict, and H_a is the guilty verdict. You assume in a jury trial that the defendant isn't guilty unless the prosecution can show beyond a reasonable doubt that he or she is guilty. If the jury says the evidence is beyond a reasonable doubt, they reject H_o, not guilty, in favor of H_a, guilty.

In general, when hypothesis testing, you set up H_o and H_a so that you believe H_o is true unless your evidence (your data and statistics) show you otherwise. And in that case, where you have sufficient evidence against H_o, you reject H_o in favor of H_a. The burden of proof is on the researcher to show sufficient evidence against H_o before it's rejected. (That's why H_a is often called the *research hypothesis*, because H_a is the hypothesis that the researcher is most interested in showing.) If H_o is rejected in favor of H_a, the researcher can say he or she has found a *statistically significant* result; that is, the results refute the previous claim, and something different or new is happening.

Finding sample statistics

After you select your sample, the appropriate number-crunching takes place. Your null hypothesis makes a statement about what the population parameter is (for example, the proportion of all women who have varicose veins or the average miles per gallon of a U.S.-built light truck). You need a measure of how much your results can be expected to change if you took a different sample. In statistical jargon, the data you collect measure that variable of interest, and the statistics that you calculate will include the sample statistic that most closely estimates the population parameter. If you're testing a claim about the proportion of women with varicose veins, you need to calculate the proportion of women in your sample who have varicose veins. If you're testing a claim about the average miles per gallon of a U.S.-built light truck, your statistic should be the average miles per gallon of the light trucks in your sample.

Standardizing the evidence: The test statistic

After you have your sample statistic, you may think you're done with the analysis part and are ready to make your conclusions — but you're not. The problem is you have no way to put your results into any kind of perspective just by looking at them in their regular units. The number of standard errors that a statistic lies above or below the mean is called a *standard score*. To interpret your statistic, you need to convert it from original units to a standard score.

When finding a standard score for a sample mean or proportion, you take your statistic, subtract the mean, and divide the result by the standard error. In the case of hypothesis tests, you use the value in H_o as the mean. (That's because you assume H_o is true, unless you have enough evidence against it.) This standardized version of your statistic is called a *test statistic,* and it's the main component of a hypothesis test.

The general procedure for converting a statistic to a test statistic (standard score):

1. **Take your statistic minus the claimed value (given by H_o).**
2. **Divide by the standard error of the statistic (see Chapter 6).**

Your test statistic represents the distance between your actual sample results and the claimed population value, in terms of number of standard errors. If you see that the distance between the claim and the sample statistic is small in terms of standard errors, your sample isn't far from the claim and your data are telling you to stick with H_o. If that distance is large, however, your data are showing less and less support for H_o. The next question is, how large of a distance is large enough to reject H_o?

Weighing the evidence and making decisions: p-values

To test whether the claim is true, you're looking at your test statistic taken from your sample, and seeing whether it supports the claim. And how do you determine that? By looking at where your test statistic ends up on its corresponding sampling distribution — see Chapter 6. In the case of means or proportions (if certain conditions are met), you look at where your test statistic ends up on the standard normal (Z) distribution. The Z-distribution has a mean of 0 and a standard deviation of 1. If your test statistic is close to 0, or at least within that range where most of the results should fall, then you can't reject the claim (H_o).

If your test statistic is out in the tails of the standard normal distribution, far from 0, it means the results of this sample do not verify the claim, hence you reject H_o. But how far is "too far from 0"? If the null hypothesis is true, most (about 95%) of the samples will result in test statistics that lie roughly within 2 standard errors of the claim. If H_a is the not-equal-to alternative, any test statistic outside this range will result in H_o being rejected (see Figure 8-1).

If your test statistic is close to 0, you can't reject the claim shown in H_o. However, this does not mean you accept the claim as truth either. Because H_o is on trial, and the test statistic is the evidence, either there is enough evidence to reject H_o or there isn't. In a real trial, the jury's conclusion is either guilty or not guilty. They never conclude "innocent." Similarly, in a hypothesis test we either say "reject H_o" or "fail to reject H_o" — we never say "accept H_o."

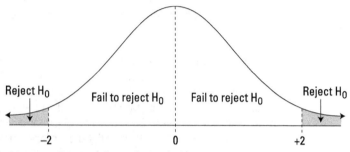

FIGURE 8-1: Test statistics and your decision.

Finding the p-value

You can be more specific about your conclusion by noting exactly how far out on the standard normal distribution the test statistic falls, so everyone knows where the result stands and what that means in terms of how strong the evidence is against the claim. In the case of means or proportions (if certain conditions are met), you do this by looking up the test statistic on the standard normal distribution (Z-distribution, Table A-1 in the appendix) and finding the probability of being at that value or beyond it (in the same direction). This p-value measures how likely it was that you would have gotten your sample results if the null hypothesis were true. The farther out your test statistic is on the tails of the standard normal distribution, the smaller the p-value will be, and the more evidence you have against the null hypothesis being true.

To find the p-value for your test statistic:

1. **Look up the location of your test statistic on the standard normal distribution (see Table A-1 in the appendix).**

2. **Find the percentage chance of being at or beyond that value in the same direction:**

   a. If H_a contains a less-than alternative (left tail), find the probability from Table A-1 in the appendix that corresponds to your test statistic.

b. If H_a contains a greater-than alternative (right tail), find the probability from Table A-1 in the appendix that corresponds to your test statistic, and then take 1 minus that. (You want the percentage to the right of your test statistic in this case, and percentiles give you the percentage to the left. See Chapter 2.)

3. **Double this probability if (and only if) H_a is the not-equal-to alternative.**

This accounts for both the less-than and the greater-than possibilities.

4. **Change the probability to a percentage by multiplying by 100 or moving the decimal point two places to the right.**

Interpreting a p-value

To make a proper decision about whether or not to reject H_o, you determine your cutoff probability for your *p*-value before doing a hypothesis test; this cutoff is called an *alpha level* (α). Typical values for α are 0.05 or 0.01. Here's how to interpret your results for any given alpha level:

» If the *p*-value is greater than or equal to α, you fail to reject H_o.

» If the *p*-value is less than α, reject H_o.

» *p*-values on the borderline (very close to α) are treated as marginal results.

Here's how you interpret your results if you use an alpha level of 0.05:

» If the *p*-value is less than 0.01 (very small), the results are considered highly statistically significant — reject H_o.

» If the *p*-value is between 0.05 and 0.01 (but not close to 0.05), the results are considered statistically significant — reject H_o.

» If the *p*-value is close to 0.05, the results are considered marginally significant — decision could go either way.

» If the *p*-value is greater than (but not close to) 0.05, the results are considered non-significant — don't reject H_o.

When you hear about a result that has been found to be statistically significant, ask for the p-value and make your own decision. Alpha levels and resulting decisions will vary from researcher to researcher.

General steps for a hypothesis test

Here's a boiled-down summary of the steps involved in doing a hypothesis test. (Particular formulas needed to find test statistics for any of the most common hypothesis tests are provided in the rest of this chapter.)

1. **Set up the null and alternative hypotheses: H_o and H_a.**

2. **Take a random sample of individuals from the population and calculate the sample statistics (means and standard deviations).**

3. **Convert the sample statistic to a test statistic by changing it to a standard score (all formulas for test statistics are provided later in this chapter).**

4. **Find the p-value for your test statistic.**

5. **Examine your p-value and make your decision.**

Testing One Population Mean

This test is used when the variable is numerical and only one population or group is being studied. For example, Dr. Phil says that the average time that working mothers spend talking to their children is 11 minutes per day. The variable, time, is numerical, and the population is all working mothers.

The null hypothesis in the Dr. Phil example is $H_o : \mu = 11$ minutes. Note that μ represents the average number of minutes per day that all working mothers spend talking to their children, and the claim is that that mean is 11. The alternative hypothesis, H_a, is either: $\mu > 11, \mu < 11$, or $\mu \neq 11$. Let's suppose you suspect that the average time working mothers spend talking with their kids is more than 11 minutes; your alternative hypothesis would be $H_a : \mu > 11$.

The formula for the test statistic for one population mean is $Z = \dfrac{\bar{x} - \mu_0}{\sigma / \sqrt{n}}$. To calculate it, do the following:

1. **Calculate the sample mean, $\bar{x}$, and the sample standard deviation, s. Let n represent the sample size.**

See Chapter 1 for calculations of the mean and standard deviation.

2. **Find σ minus μ_o. (Remember, μ_o is the claimed value of the population mean.)**

3. **Calculate the standard error: $\sigma / \sqrt{n}$.**

4. **Divide your result from Step 2 by the standard error found in Step 3.**

For the Dr. Phil example, suppose a random sample of 100 working mothers spend an average of 11.5 minutes per day talking with their children, with a standard deviation of 2.3 minutes. That means $\bar{x}$ is 11.5, where $n = 100$ and $s = 2.3$. Take $11.5 - 11 = +0.5$. Take 2.3 divided by the square root of 100 (which is 10) to get 0.23 for the standard error. Divide +0.5 by 0.23, to get 2.17. That's your test statistic.

This means your sample mean is 2.17 standard errors above the claimed population mean. Would these sample results be unusual if the claim ($H_o : \mu = 11$ minutes) were true? To decide whether your test statistic supports H_o, calculate the p-value. To calculate the p-value, look up your test statistic (in this case, 2.17) on the standard normal distribution (Z-distribution) — see Table A-1 in the appendix — and take 100% minus the percentile shown (since you are looking at the right tail), because your H_a is a greater-than hypothesis. In this case, the percentage would be $100\% - 98.50\% = 1.50\%$. So, the p-value is 0.0150 (1.50%).

This p-value of 0.0139 (1.39%) is much less than 0.05 (5%). So, reject the claim ($\mu = 11$ minutes) by rejecting H_o, and concluding H_a ($\mu > 11$ minutes). Your conclusion: According to this (hypothetical) sample, Dr. Phil's claim of 11 minutes is rejected; the actual average is greater than 11 minutes per day.

TECHNICAL STUFF

If the sample size, n, were less than 30 here, or the population standard deviation, σ, were unknown, you would look up your test statistic on the t-distribution with $n - 1$ degrees of freedom (see Chapter 9) rather than the Z-distribution.

Testing One Population Proportion

This test is used when the variable is categorical (for example, gender or political party) and only one population is being studied (for example, all U.S. citizens). The test is looking at the proportion (p) of individuals in the population who have a certain characteristic — for example, the proportion of people who carry cellphones. The null hypothesis is $H_o : p = p_o$, where p_o is a certain claimed value. For example, if the claim is 20% of people carry cellphones, p_o is 0.20. The alternative hypothesis is one of the following: $p > p_o, p < p_o,$ or $p \neq p_o$.

The formula for the test statistic for a single proportion is $\dfrac{\hat{p} - p_o}{\sqrt{\dfrac{p_o(1-p_o)}{n}}}$. To calculate it, do the following:

1. Calculate the sample proportion, $\hat{p}$, by taking the number of people in the sample who have the characteristic of interest (for example, the number of people in the sample carrying cellphones) and dividing that by n, the sample size.

2. Take $\hat{p}$ minus p_o. (Remember p_o is the claimed number for the population proportion.)

3. Calculate the standard error: $\sqrt{\dfrac{p_o(1-p_o)}{n}}$.

4. Divide your result from Step 2 by your result from Step 3.

To interpret the test statistic, look up your test statistic on the standard normal distribution (see Table A-1 in the appendix) and calculate the p-value. For example, suppose Cavifree toothpaste claims that four out of five dentists recommend Cavifree toothpaste to their patients. In this case, the population is all dentists, and p is the proportion of all dentists who recommended Cavifree to their patients. The claim is that p is equal to "four out of five," which means that p_o is $4/5 = 0.80$. You suspect that the proportion is actually less than 0.80. Your hypotheses are $H_o : p = 0.80$ versus $H_a : p < 0.80$. Suppose that 150 out of 200 dental patients sampled received a recommendation for Cavifree.

To find the test statistic, observe that the sample proportion $\hat{p}$ is $150/200 = 0.75$. Since $p_o = 0.80$, take $0.75 - 0.80 = -0.05$ as your numerator. Next, the standard error is the square root of $[(0.80 * [1 - 0.80]) / 200] =$ the square root of $(0.16 / 200) =$ the square root of $0.0008 = 0.028$. The test statistic is -0.05 divided by 0.028, which is $-0.05 / 0.028 = -1.79$. This means that your sample results are 1.79 standard errors below the claimed value for the population.

How often would you expect to get results like this if H_o were true? The percentage chance of being at or beyond (in this case to the left of) -1.79, is 3.67%. (Look up -1.79 in Table A-1 in the appendix and use the corresponding percentile, because H_a is a less-than hypothesis.) Now divide by 100 to get your p-value, which is 0.0367. Because the p-value is less than 0.05, you have enough evidence to reject H_o. According to your sample, the claim of four out of five (80% of) dentists recommending Cavifree toothpaste is not true; the actual percentage of recommendations is less than that.

Comparing Two Population Means

This test is used when the variable is numerical (for example, income, cholesterol level, or miles per gallon) and two populations or groups are being compared (for example, cars versus SUVs). Two separate random samples need to be selected, one from each population, in order to collect the data needed for this test. The null hypothesis is that the two population means are the same; in other words, that their difference is equal to 0. The notation for the null hypothesis is $H_o : \mu_x - \mu_y = 0$, where μ_x is the mean of the first population, and μ_y is the mean of the second population.

The test statistic comparing two means is:

$$\frac{(\bar{x} - \bar{y}) - 0}{\sqrt{\dfrac{s_x^2}{n_1} + \dfrac{s_y^2}{n_2}}}$$

To calculate it, do the following:

1. **Calculate the sample means ($\bar{x}$ and $\bar{y}$) and sample standard deviations (s_x and s_y) for each sample separately. Let n_1 and n_2 represent the two sample sizes (they need not be equal).**

   See Chapter 1 for these calculations.

2. **Find the difference between the two sample means,** $\bar{x} - \bar{y}$.

3. **Calculate the standard error,** $\sqrt{\dfrac{s_x^2}{n_1} + \dfrac{s_y^2}{n_2}}$.

4. **Divide your result from Step 2 by your result from Step 3.**

To interpret the test statistic, look up your test statistic on the t-distribution with $n_1 + n_2 - 2$ degrees of freedom (see Table A-2 in the appendix) and calculate the p-value. For example, suppose you want to compare the absorbency of two brands of paper towels (call the brands Stats-absorbent and Sponge-o-matic). You can make this comparison by looking at the average number of ounces each brand can absorb before being saturated. H_o says the difference between the average absorbencies is 0 (non-existent), and H_a says the difference is not 0. In other words, $H_o : \mu_x - \mu_y = 0$ versus $H_o : \mu_x - \mu_y \neq 0$. Here, you have no indication of which paper towel may be more absorbent, so the not-equal-to alternative is the one to use.

Suppose you select a random sample of 50 paper towels from each brand and measure the absorbency of each paper towel. Suppose the average absorbency of Stats-absorbent (x) is 3 ounces, with a standard deviation of 0.9 ounces, and for Sponge-o-matic (y), the average absorbency is 3.5 ounces, with a standard deviation of 1.2 ounces.

Given these data, you have $\bar{x} = 3, s_x = 0.9, \bar{y} = 3.5, s_y = 1.2, n_1 = 50$, and $n_2 = 50$. The difference between the sample means for (Stats-absorbent − Sponge-o-matic) is $(3 - 3.5) = -0.5$ ounces. (A negative difference simply means that the second sample mean was larger than the first.) The standard error is $\sqrt{\dfrac{0.9^2}{50} + \dfrac{1.2^2}{50}} = \sqrt{\dfrac{0.81}{50} + \dfrac{1.44}{50}} = .2121$. Divide the difference, −0.5, by the standard error, 0.2121, which gives you −2.36. This is your test statistic.

To find the p-value, look up −2.36 on the Z-table (Table A-1 in the appendix). The chance of being beyond, in this case to the left of, −2.36 is equal to the percentile, which is 0.91%. Because H_a is a not-equal-to alternative, you double this percentage to get $2 \times 0.91\% = 1.82\%$. Change this to a probability by dividing by 100 to get a p-value of 0.0182. This p-value is less than 0.05. That means you do have enough evidence to reject H_o.

Your conclusion is that a statistically significant difference exists between the absorbency levels of these two brands of paper towels, based on your samples. Sponge-o-matic comes out on top because it has a higher average.

TECHNICAL STUFF

If either of the sample sizes is small (generally less than 30), you use the t-distribution with $n_1 + n_2 - 2$ degrees of freedom (see Chapter 9) instead of the standard normal distribution when figuring out the p-value.

Testing the Mean Difference: Paired Data

This test is used when the variable is numerical (for example, cholesterol level or miles per gallon), and the individuals in the sample are either paired up in some way (identical twins are often used) or the same people are used twice (for example, using a pre-test and post-test). Paired tests are used for comparisons where you want to minimize the chance of the treatment and control groups being too different (and hence biased). See Chapter 13 for details.

Suppose a researcher wants to see whether teaching students to read using a computer game gives better results than teaching with a tried-and-true phonics method. She randomly selects 20 students and puts them into 10 pairs according to their reading readiness level, age, IQ, and so on. She randomly selects one student from each pair to learn to read via the computer game, and the other learns to read using the phonics method. At the end of the study, each student takes the same reading test. The data are shown in Table 8-1.

The data are in pairs, but you're really interested only in the difference in reading scores (computer reading score – phonics reading score) for each pair, not the reading scores themselves. So, you take the difference between the scores for each pair, and those *paired differences* make up your new set of data to work with. If the two reading methods are the same, the average of the paired differences should be 0. If the computer method is better, the average of the paired differences should be positive (because the computer reading score should be larger than the phonics score).

TABLE 8-1 **Reading Scores for Computer Game versus the Phonics Method**

Student Pair #	Reading Score for Computer Method	Reading Score for Phonics Method	Paired Differences (Computer Score – Phonics Score)
1	85	80	+5
2	80	80	+0
3	95	88	+7
4	87	90	–3
5	78	72	+6
6	82	79	+3
7	57	50	+7
8	69	73	–4
9	73	78	–5
10	99	95	+4

Testing paired data amounts to testing one population mean, where the null hypothesis is that the mean (of the paired differences) is 0, and the alternative hypothesis is that the mean (of the paired differences) is > 0; < 0, or $\neq 0$. The notation for the null hypothesis is $H_o : \mu_d = 0$, where μ_d is the population mean of all paired differences. (The d in the subscript reminds you that you're working with the paired differences.)

The formula for the test statistic for paired differences is $\dfrac{\bar{d} - \mu_d}{s_d / \sqrt{n}}$. To calculate it, do the following:

1. **For each pair of data, take the first value in the pair minus the second value in the pair to find the paired difference.**

   Think of the differences as your new data set.

2. **Calculate the mean, $\bar{d}$, and the standard deviation, s_d, of all the differences in the pairs in the sample.**

Let n represent the number of paired differences that you have.

3. **Calculate the standard error:** $s_d\big/\sqrt{n}$.

4. **Take $\bar{d}$ divided by the standard error from Step 3.**

Remember that $\mu_d = 0$ if H_o is true, so it's not included in the formula here.

For the reading scores example, you can use these steps to see whether the computer method is better at teaching students to read. Calculate the differences for each pair; you can see those differences in column 4 of Table 8-1. Notice that the sign on each of the differences is important; it indicates which method performed better for that particular pair.

The mean and standard deviation of the differences (column 4 of Table 8-1) must be calculated. The mean of the differences is found to be +2, and the standard deviation is 4.64. Note that $n = 10$ here. The standard error is 4.64 divided by the square root of 10 (which is 3.16). So you have $4.64 / 3.16 = 1.47$. (Remember that n is the number of pairs, which is 10.) For the last step, take the mean of the differences, +2, divided by the standard error, which is 1.47, to get +1.36, the test statistic. That means the average difference for this sample is 1.36 standard errors above 0. Is this enough to say that a difference in reading scores applies to the whole population?

Because n is less than 30, you look up 1.36 on the t-distribution with $10 - 1 = 9$ degrees of freedom (see Table A-2 in the appendix) to calculate the p-value (see Chapter 9). The p-value in this case is greater than 0.05 because 1.36 is close to the value of 1.38 on the table, and, therefore its p-value would be about 0.10 (the corresponding p-value for 1.38). That's because 1.38 is in the column under the 90th percentile, and because H_a is a greater-than alternative, you take $100\% - 90\% = 10\% = 0.10$. Since the p-value is clearly greater than 0.05, you conclude that there isn't enough evidence to reject H_o, so the computer game can't be touted as a better reading method. (This could be due to the lack of additional evidence needed to prove this with a smaller sample size.)

TECHNICAL
STUFF

In many paired experiments, the data sets will be small due to costs and time associated with doing these kinds of studies. That means the t-distribution with $n-1$ degrees of freedom (see Chapter 9) is often used instead of the standard normal distribution (see Table A-1 in the appendix) when figuring out the p-value.

Testing Two Population Proportions

This test is used when the variable is categorical (for example, smoker/nonsmoker, political party, support/oppose an opinion, and so on) and you're interested in the proportion of individuals with a certain characteristic — for example, the proportion of smokers. In this case, two populations or groups are being compared (such as the proportion of female smokers versus male smokers).

In order to conduct this test, two separate random samples need to be selected, one from each population. The null hypothesis is that the two population proportions are the same; in other words, that their difference is equal to 0. The notation for the null hypothesis is $H_o : p_1 - p_2 = 0$, where p_1 is the proportion from the first population, and p_2 is the proportion from the second population.

Here is the formula for the test statistic comparing two proportions:

$$\frac{(\hat{p}_1 - \hat{p}_2) - 0}{\sqrt{\hat{p}(1-\hat{p})\left(\frac{1}{n_1} + \frac{1}{n_2}\right)}}$$

where $\hat{p}$ is the *pooled sample proportion*, also known as the proportion of all individuals from the combined samples that have the characteristic of interest. To calculate it, do the following:

1. **Calculate the sample proportions $\hat{p}_1$ and $\hat{p}_2$.**

   For each sample, let n_1 and n_2 represent the two sample sizes (they need not be equal).

2. **Find the difference between the two sample proportions, $(\hat{p}_1 - \hat{p}_2)$.**

3. Calculate the pooled sample proportion, $\hat{p}$, which is the total number of individuals from both samples who have the characteristic of interest (for example, the total number of smokers, male or female, in the sample), divided by the total number of individuals from both samples $(n_1 + n_2)$.

4. Calculate the standard error:

$$\sqrt{\hat{p} \cdot (1-\hat{p})\left(\frac{1}{n_1} + \frac{1}{n_2}\right)}$$

5. Divide your result from Step 2 by your result from Step 4.

   To interpret the test statistic, look up your test statistic on the standard normal distribution (Table A-1 in the appendix) and calculate the p-value.

For example the maker of Adderall, a drug for attention deficit hyperactivity disorder (ADHD), reported that 26 of the 374 subjects (7%) who took the drug experienced vomiting as a side effect, compared to 8 of the 210 subjects (4%) who were on a *placebo* (fake drug). Note that patients didn't know which treatment they were given. In the sample, more people on the drug experienced vomiting, but is this percentage enough to say that the entire population would experience more vomiting? You can test it to see. In this case, you have $H_o : p_1 - p_2 = 0$ versus $H_a : p_1 - p_2 > 0$, where p_1 represents the proportion of subjects who vomited using Adderall, and p_2 represents the proportion of subjects who vomited using the placebo.

TIP

Why does H_a contain a ">" sign and not a "<" sign? H_a represents the scenario in which those taking Adderall experience more vomiting than those on placebo — that's something the FDA would want to know about.

The next step is calculating the test statistic. First, $\bar{p}_1 = 26/374 = 0.07$ and $\bar{p}_2 = 8/210 = 0.04$. The sample sizes are $n_1 = 374$ and $n_2 = 210$, respectively. Next, take the difference between these sample proportions to get $0.07 - 0.04 = 0.03$. The overall sample proportion, $\hat{p}$, is $(26+8)/(374+210) = 34/584 = 0.058$. The standard error is $\sqrt{0.058(1-0.058)\left(\frac{1}{374}\right)+\left(\frac{1}{210}\right)} = 0.02$. Finally, take the difference from Step 2, 0.03, divided by 0.02 to get $0.03/0.02 = 1.5$, which is the test statistic.

The p-value is the percentage chance of being at or beyond (in this case, to the right of) 1.5, which is $100\% - 93.32\% = 6.68\%$, which is written as a probability of 0.0668. This p-value is greater than 0.05, so you don't have enough evidence to reject H_o. That means vomiting is not experienced any more by those taking this drug when compared to those taking a placebo.

You Could Be Wrong: Errors in Hypothesis Testing

After you decide whether to reject H_o, the next step is living with the consequences — after all, you could be wrong.

>> If you conclude that a claim isn't true but it actually *is* true, a lawsuit, fine, unnecessary changes in the product, or consumer boycotts that shouldn't have happened could result.

>> If you conclude that a claim is true but it actually isn't, what happens then? Undetected problems will continue and no action will be taken. Inaction has consequences as well.

Rejecting H_o when you shouldn't is called a *Type-1 error*. I don't really like this name, because it seems so nondescript. I prefer to call a Type-1 error a *false alarm*. In the case of the packages, if the consumer group made a Type-1 error when it rejected the company's claim, they created a false alarm. What's the result? A very angry delivery company.

A false alarm: Type-1 error

Suppose a company claims that its average package delivery time is 2 days, and a consumer group tests this hypothesis and concludes that the claim is false: They believe that the average delivery time is actually more than 2 days. This is a big deal. If the group can stand by its statistics, it has done well to inform the public about the false advertising issue. But what if the group is wrong? Even if the study is based on a good design, collects good data, and makes the right analysis, the group can still be wrong.

Why? Because its conclusions were based on a sample of packages, not on the entire population. As Chapter 6 tells you, sample results vary from sample to sample. If your test statistic falls on the tail of the standard normal distribution, these results are unusual, if the claim is true, because you expect them to be much closer to the middle of the standard normal distribution (Z-distribution). Just because the results from a sample are unusual, however, doesn't mean they're impossible. A p-value of 0.04 means that the chance of getting your particular test statistic (out on the tail of the standard normal distribution), even if the claim is true, is 4% (less than 5%). That's why you reject H_o in this case, because that chance is so small. But a chance is a chance!

Perhaps your sample, while collected randomly, just happens to be one of those atypical samples whose result ended up far out on the distribution. So H_o could be true, but your results lead you to a different conclusion. How often does that happen? Five percent of the time (or whatever your given alpha level is for rejecting H_o).

A missed detection: Type-2 error

Now suppose the company really wasn't delivering on its claim. Who's to say that the consumer group's sample will detect it? If the actual delivery time is 2.1 days instead of 2 days, the difference would be pretty hard to detect. If the actual delivery time is 3 days, a fairly small sample would show that something's up. The issue lies with those in-between values, like 2.5 days. If H_o is indeed false, you want to detect that and reject H_o. Not rejecting H_o when you should have is called a *Type-2 error*. I call it a *missed detection*.

Sample size is the key to being able to detect situations where H_o is false and to avoiding Type-2 errors. The more information you have, the less variable your results will be, and easier it will be to detect problems that exist with a claim.

This ability to detect when H_o is truly false is called the *power* of a test. Power is a pretty complicated issue, but what's important for you to know is that the higher the sample size, the more powerful a test is. A powerful test has a small chance for a Type-2 error.

Statisticians recommend two preventative measures to minimize the chances of a Type-1 or Type-2 error:

» Set a low cutoff probability for rejecting H_o (like 5 percent or 1 percent) to reduce the chance of false alarms (minimizing Type-1 errors).

» Select a large sample size to ensure that any differences or departures that really exist won't be missed (minimizing Type-2 errors).

Chapter **9**
The *t*-Distribution

Many different distributions exist in statistics, and one of the most commonly used distributions is the *t*-distribution. In this chapter, I go over the basic characteristics of the *t*-distribution, how to use the *t*-table to find probabilities, and how it's used to solve problems in its most well-known settings — confidence intervals and hypothesis tests.

Basics of the t-Distribution

The normal distribution is the well-known bell-shaped distribution whose mean is μ and whose standard deviation is σ. (See Chapter 5 for more on normal distributions.) The *t*-distribution can be thought of as a cousin of the normal distribution — it looks similar to a normal distribution in that it has a basic bell shape with an area of 1 under it, but is shorter and flatter than a normal distribution. Like the standard normal (Z) distribution, it is centered at zero, but its standard deviation is proportionally larger compared to the *Z*-distribution.

As with normal distributions, there is an entire family of different *t*-distributions. Each *t*-distribution is distinguished by what statisticians call *degrees of freedom*, which are related to the sample size of the data set. If your sample size is *n*, the degrees of freedom for the corresponding *t*-distribution is $n-1$. For example, if your sample size is 10, you use a t-distribution with $10-1$ or 9 degrees of freedom, denoted t_9.

Smaller sample sizes have flatter t-distributions than larger sample sizes. And as you may expect, the larger the sample size is, and the larger the degrees of freedom, the more the t-distribution looks like a standard normal distribution (Z-distribution); and the point where they become very similar (similar enough for jazz or government work) is about the point where the sample size is 30. (This result is due to the Central Limit Theorem; see Chapter 6.)

Figure 9-1 shows different t-distributions for different sample sizes, and how they compare to the Z-distribution.

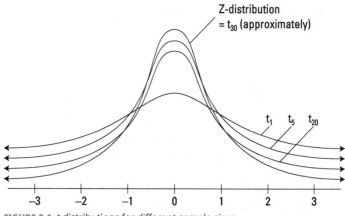

FIGURE 9-1: t-distributions for different sample sizes.

Understanding the t-Table

Each t-distribution has its own shape and its own set of probabilities, so one size doesn't fit all. To help with this, statisticians have come up with one abbreviated table that you can use to mark off certain points of interest on several different t-distributions whose degrees of freedom range from 1 to 30 (see Appendix Table A-2). If you look at the column headings in Table A-2, you see selected values from 0.40 to 0.0005. These numbers represent right tail probabilities (the probability of being larger than a certain value). The numbers moving down any given column represent the values on each t-distribution having those right tail probabilities. For example, under the "0.05" column, the first number is 6.313752. This represents the number on the t_1 distribution (one degree of freedom) whose probability to the right equals 0.05. Farther down that column in row 15 you see 1.753050. This is the number on the t_{15} distribution whose probability to the right is 0.05.

You can also use the t-table to find percentiles for the t-distribution (recall that a percentile is a number whose area to the *left* is a given percentage). For example, suppose you have a sample of size 10 and you want to find the 95th percentile. You have $n-1=9$ degrees of freedom, so you look at the row for 9, and the column for 0.05 to get $t=1.833$. Since the area to the right of 1.833 is 0.05, that means the area to the left must be $1-0.05=0.95$ or 95%. You have found the 95th percentile of the t_9 distribution: 1.833. Now, if you increase the sample size to $n=20$, the 95th percentile decreases; look at the row for $20-1=19$ degrees of freedom, and in the 0.05 column you find $t=1.729$. Remember as n gets large, the values on a t-distribution are more condensed around the mean, giving it a more curved shape, like the Z-distribution.

Notice that as the degrees of freedom of the t-distribution increase (as you move down any given column in Table A-2 in the appendix), the t-values get smaller and smaller. The last row of the table corresponds to the values on the Z-distribution. That confirms what you already know: As the sample size increases, the t- and the Z-distributions are more and more alike. The degrees of freedom for the last row of the t-table (Table A-2) are listed as "infinity" to make the point that a t-distribution approaches a Z-distribution as n gets infinitely large.

t-Distributions and Hypothesis Tests

The most common use by far of the t-distribution is in hypothesis testing — in particular, the case where you do a hypothesis test for one population mean. (See Chapter 8 for the whole scoop on hypothesis testing.) You use a t-distribution when you do not know the standard deviation of the population (σ) and you have to use the standard deviation of the sample (s) to estimate it. Typically in these situations you also have a small sample size (but not always).

Finding critical values

If you don't know the population standard deviation and are using the sample standard deviation instead, you pay a penalty: a t-distribution with more variability and fatter tails. When you do a hypothesis test, your "cutoff point" for rejecting the null hypothesis H_0 is further out than it would have been if you had more data and could use the Z-distribution. (For more information on hypothesis tests and how they are set up, see Chapter 8.)

For example, suppose you have a two-tailed hypothesis test for one population mean where $\alpha = 0.05$ and you have a sample size of 100. If you are doing a two-sided hypothesis test for one population mean, you can use $Z =$ plus or minus 1.96 as your critical value to determine whether to reject H_0. But if n is, say, 8, the critical value for this same test would be $t_7 =$ plus or minus 2.365; see Table A-2, row 7, column "0.025" to obtain this number. (Remember a two-tailed hypothesis test with significance level 5% has $2.5\% = 0.025$ in each tail area.) This means you have to submit more evidence to reject H_0 if you only have 8 pieces of data than if you have 100 pieces of data. In other words, your goal line to make a touchdown and find a "statistically significant result" is set at 2.365 for the t-distribution versus 1.96 for the Z-distribution.

Finding p-values

Recall from hypothesis testing (Chapter 8) that a p-value is the probability of obtaining a result beyond your test statistic on the appropriate distribution. In terms of p-values, the same test statistic has a larger p-value on a t-distribution than on the Z-distribution. A test statistic far out on the leaner Z-distribution has little area beyond it. But that same test statistic out on the fatter t-distribution has more fat (or area) beyond it, and that's exactly what the p-value represents.

Suppose your sample size is 10, your test statistic (referred to as the t-value) is 2.5, and your alternative hypothesis, H_a, is the greater-than alternative. Because the sample size is 10, you use the t-distribution with $10-1=9$ degrees of freedom to calculate your p-value. This means you'll be looking at the row in the t-table (Table A-2 in the appendix) that has a 9 in the Degrees of Freedom column. Your test statistic (2.5) falls between two values: 2.262 (the "0.025" column) and 2.821 (the "0.01" column).

The p-value is between $0.025 = 2.5\%$ and $0.01 = 1\%$. You don't know exactly what the p-value is, but because 1% and 2.5 % are both less than the typical cutoff of 5%, you reject H_o.

TECHNICAL STUFF

The t-table (Table A-2 in the appendix) doesn't include all possible test statistics on it, so simply choose the one test statistic that's closest to yours, look at the column it's in, and find the corresponding percentile. Then figure your p-value.

Note that for a less-than alternative hypothesis, your test statistic would be a negative number (to the left of 0 on the t-distribution). In this case, you want to find the percentage below, or to the left of, your test statistic to get your p-value. Yet negative test statistics don't appear on Table A-2. Not to worry! The percentage to the left (below) a negative t-value is the same as the percentage to the right (above) the positive t-value, due to symmetry. So, to find the p-value for your negative test statistic, look up the positive version of your test statistic in Table A-2, and find the corresponding right tail probability. For example, if your test statistic is –2.5 with 9 degrees of freedom, look up +2.5 on Table A-2, and you find that it falls between the 0.025 and 0.01 columns, so your p-value is somewhere between 1% and 2.5%.

If your alternative hypothesis (H_a) has the not-equal-to alternative, double the percentage that you get to obtain your p-value. That's because the p-value is this case represents the chance of being beyond your test statistic in either the positive or negative direction (see Chapter 8 for details on hypothesis testing).

t-Distributions and Confidence Intervals

The t-distribution is used in a similar way with confidence intervals (see Chapter 7 for more on confidence intervals). If your data have a normal distribution and either 1) the sample size is small; or 2) you don't know the population standard deviation, σ, and you must use the sample standard deviation, s, to substitute for it, you use a value from a t-distribution with $n-1$ degrees of freedom instead of a z-value in your formulas for the confidence intervals for the population mean.

For example, to make a 95% confidence interval for μ where $n = 9$ you add and subtract 2.306 times the standard error of $\bar{X}$ when you use the t-distribution versus adding and subtracting 1.96 times the standard error when you use a z-value.

Chapter **10**
Correlation and Regression

I n this chapter, you analyze two numerical variables, X and Y, to look for patterns, find the correlation, and make predictions about Y from X, if appropriate, using simple linear regression.

Picturing the Relationship with a Scatterplot

A fair amount of research supports the claim that the frequency of cricket chirps is related to temperature. And this relationship is actually used at times to predict the temperature using the number of times the crickets chirp per 15 seconds. To illustrate, I've taken a subset of some of the data that's been collected on this; you can see it in Table 10-1.

Notice that each observation is composed of two variables that are tied together, in this case the number of times the cricket chirped in 15 seconds (the X-variable), and the temperature at the time the data was collected (the Y-variable). Statisticians call this type of two-dimensional data *bivariate* data. Each observation contains one pair of data collected simultaneously.

TABLE 10-1 Cricket Chirps and Temperature Data (Excerpt)

Number of Chirps (in 15 Seconds)	Temperature (Fahrenheit)
18	57
20	60
21	64
23	65
27	68
30	71
34	74
39	77

Making a scatterplot

Bivariate data are typically organized in a graph that statisticians call a *scatterplot*. A scatterplot has two dimensions, a horizontal dimension (called the x-axis) and a vertical dimension (called the y-axis). Both axes are numerical — each contains a number line.

The x-coordinate of bivariate data corresponds to the first piece of data in the pair; the y-coordinate corresponds to the second piece of data in the pair. If you intersect the two coordinates, you can graph the pair of data on a scatterplot. Figure 10-1 shows a scatterplot of the data from Table 10-1.

Interpreting a scatterplot

You interpret a scatterplot by looking for trends in the data as you go from left to right:

>> If the data show an uphill pattern as you move from left to right, this indicates a *positive relationship between X and Y*. As the x-values increase (move right), the y-values increase (move up) a certain amount.

>> If the data show a downhill pattern as you move from left to right, this indicates a *negative relationship between X and Y*. That means as the x-values increase (move right), the y-values decrease (move down) by a certain amount.

>> If the data don't resemble any kind of pattern (even a vague one), then no relationship exists between *X* and *Y*.

This chapter focuses on linear relationships. A *linear relationship between X and Y* exists when the pattern of *x*- and *y*-values resembles a line, either uphill (with positive slope) or downhill (with negative slope).

Looking at Figure 10-1, there does appear to be a positive linear relationship between number of cricket chirps and the temperature. That is, as the cricket chirps increase, you can predict that the temperature is higher as well.

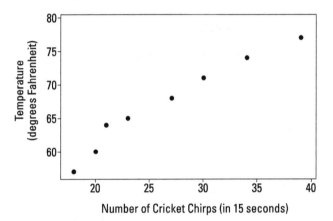

FIGURE 10-1: Scatterplot of cricket chirps versus outdoor temperature.

Measuring Relationships Using the Correlation

After the bivariate data have been organized, the next step is to do some statistics that can quantify or measure the extent and nature of the relationship.

Calculating the correlation

The pattern and direction of the relationship between *X* and *Y* can be seen from the scatterplot. The strength of the relationship between two numerical variables depends on how closely the data resemble a certain pattern. Although many different types of patterns can exist between two variables, this chapter examines linear patterns only.

Statisticians use the *correlation coefficient* to measure the strength and direction of the linear relationship between two numerical variables X and Y. The correlation coefficient for a sample of data is denoted by r.

TECHNICAL STUFF

Although the street definition of *correlation* applies to any two items that are related (such as gender and political affiliation), statisticians only use this term in the context of two numerical variables. The formal term for correlation is the *correlation coefficient*. Many different correlation measures have been created; the one in our case is the Pearson correlation coefficient (I'll just call it the correlation).

The formula for the correlation (r) is

$$r = \frac{1}{n-1}\sum \frac{(x - \bar{x})(y - \bar{y})}{s_x s_y}$$

where n is the number of pairs of data; $\bar{x}$ and $\bar{y}$ are the sample means; and s_x and s_y are the sample standard deviations of the x- and y- values, respectively.

To calculate the correlation r from a data set:

1. **Find the mean of all the x-values ($\bar{x}$) and the mean of all the y-values ($\bar{y}$).**

   See Chapter 2 for information on the mean.

2. **Find the standard deviation of all the x-values (call it s_x) and the standard deviation of all the y-values (call it s_y).**

   See Chapter 2 for information on standard deviation.

3. **For each (x, y) pair in the data set, take x minus $\bar{x}$ and y minus $\bar{y}$, and multiply them together.**

4. **Add up all the results from Step 3.**

5. **Divide the sum by $s_x * s_y$.**

6. **Divide the result by $n - 1$, where n is the number of (x, y) pairs.**

   This gives you the correlation r.

For example, suppose you have the data set (3, 2), (3, 3), and (6, 4). Following the preceding steps, you can calculate the correlation coefficient r via the following steps. (Note that for this data the x-values are 3, 3, 6, and the y-values are 2, 3, 4.)

1. $\bar{x}$ is $12/3 = 4$, and $\bar{y}$ is $9/3 = 3$.

2. The standard deviations are calculated to be $s_x = 1.73$ and $s_y = 1.00$.

3. The differences found in Step 2 multiplied together are:
   $(3 - 4)(2 - 3) = (-1)(-1) = 1$; $(3 - 4)(3 - 3) = (-1)(0) = 0$; $(6 - 4)(4 - 3) = (+2)(+1) = +2$.

4. Adding the Step 3 results, you get $1 + 0 + 2 = 3$.

5. Dividing by $s_x * s_y$ gives you $3/(1.73 * 1.00) = 3/1.73 = 1.73$.

6. Now divide the Step 5 result by $3 - 1$ (which is 2) and you get the correlation $r = 0.87$.

Interpreting the correlation

The correlation r is always between $+1$ and -1. Here is how you interpret various values of r. A correlation that is

» Exactly -1 indicates a perfect downhill linear relationship.

» Close to -1 indicates a strong downhill linear relationship.

» Close to 0 means no linear relationship exists.

» Close to $+1$ indicates a strong uphill linear relationship.

» Exactly $+1$ indicates a perfect uphill linear relationship.

TECHNICAL STUFF

How "close" do you have to get to -1 or $+1$ to indicate a strong linear relationship? Most statisticians like to see correlations above $+0.60$ (or below -0.60) before getting too excited about them. Don't expect a correlation to always be $+0.99$ or -0.99; real data aren't perfect.

Figure 10-2 shows examples of what various correlations look like in terms of the strength and direction of the relationship.

For my subset of the cricket chirps versus temperature data, I calculated a correlation of 0.98, which is almost unheard of in the real world (these crickets are *good!*).

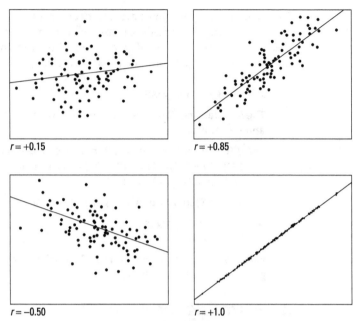

r = +0.15   r = +0.85

r = −0.50   r = +1.0

FIGURE 10-2: Scatterplots with various correlations.

Properties of the correlation

Here are two important properties of correlation:

» The correlation is a unitless measure. This means that if you change the units of X or Y, the correlation doesn't change. For example, changing the temperature (Y) from Fahrenheit to Celsius won't affect the correlation between the frequency of chirps and the outside temperature.

» The variables X and Y can be switched in the data set, and the correlation doesn't change. For example, if height and weight have a correlation of 0.53, weight and height have the same correlation.

Finding the Regression Line

After you've found a linear pattern in the scatterplot, and the correlation between the two numerical variables is moderate to strong, you can create an equation that allows you to predict one variable using the other. This equation is called the *simple linear regression line.*

Which is X and which is Y?

Before moving forward with your regression analysis, you have to identify which of your two variables is *X* and which is *Y*. When doing correlations, the choice of which variable is *X* and which is *Y* doesn't matter, as long as you're consistent for all the data; but when fitting lines and making predictions, the choice of *X* and *Y* makes a difference. In general, *X* is the variable that is the predictor. Statisticians call the *X*-variable (here cricket chirps) the *explanatory variable,* because if *X* changes, the slope tells you (or explains) how much *Y* is expected to change. The *Y*-variable (here temperature) is called the *response variable* because if *X* changes, the response (according the equation of the line) is a change in *Y*. Hence *Y* can be predicted by *X* if a strong relationship exists.

Note: In this example, I want to predict the temperature based on listening to crickets. Obviously, the real cause-and-effect is the opposite: As temperature rises, crickets chirp more.

Checking the conditions

In the case of two numerical variables, it's possible to come up with a line that you can use to predict *Y* from *X*, if (and only if) the following two conditions you examined in the previous sections are met: 1) The scatterplot must find a linear pattern; and 2) The correlation, *r*, is moderate to strong (typically beyond ±0.60).

It's not always the case that folks actually check these conditions. I've seen cases where researchers go ahead and make predictions when a correlation was as low as 0.20, or where the data follow a curve instead of a line when you make the scatterplot! That doesn't make any sense.

But suppose the correlation *is* high; do you need to look at the scatterplot? Yes. There are situations where the data have a somewhat curved shape, yet the correlation is still strong.

Understanding the equation

For the crickets and temperature data, you see the scatterplot in Figure 10-1 shows a linear pattern. The correlation between cricket chirps and temperature was found to be very strong ($r = 0.98$). You now can find one line that best fits the data (in terms of the having the smallest average distance to all the points). Statisticians call this technique for finding the best-fitting line a *simple linear regression analysis*.

Do you have to try lots of different lines to see which one fits best? Fortunately, this is not the case (although eyeballing a line on the scatterplot does help you think about what you'd expect the answer to be). The best-fitting line has a distinct slope and y-intercept that can be calculated using formulas (and, I may add, these formulas aren't too hard to calculate).

The formula for the *best-fitting line* (or *regression line*) is $y = mx + b$, where m is the slope of the line and b is the y-intercept. (This is the same equation from algebra.) The slope of a line is the change in Y over the change in X. For example, a slope of 10/3 means as the x-value increases (moves right) by 3 units, the y-value moves up by 10 units on average.

The y-intercept is that place on the y-axis where the line crosses. For example, in the equation $y = 2x - 6$, the line crosses the y-axis at the point –6. The coordinates of this point are (0,–6); when a line crosses the y-axis, the x-value is always 0. To come up with the best-fitting line, you need to find values for m and b that fit the pattern of data the absolute best. The following sections find these values.

Finding the slope

The formula for the slope, m, of the best-fitting line is $m = r\left(\dfrac{s_y}{s_x}\right)$ where r is the correlation between X and Y, and s_x and s_y are the standard deviations of the x-values and the y-values. To calculate the slope, m, of the best-fitting line:

1. **Divide s_y by s_x.**
2. **Multiply the result in Step 1 by r.**

TECHNICAL
STUFF

The correlation and the slope of the best-fitting line are not the same. The formula for slope takes the correlation (a unitless measurement) and attaches units to it. Think of s_y / s_x as the change in Y over the change in X, in units of X and Y; for example, change in temperature (degrees Fahrenheit) per increase of one cricket chirp (in 15 seconds).

Finding the y-intercept

The formula for the y-intercept, b, of the best-fitting line is $b = \bar{y} - m\bar{x}$, where $\bar{x}$ and $\bar{y}$ are the means of the x-values and the y-values, respectively, and m is the slope (the formula for which is given in the preceding section). To calculate the y-intercept, b, of the best-fitting line:

1. **Find the slope, *m*, of the best-fitting line using the steps listed in the preceding section.**

2. **Multiply by $\bar{x}$.**

3. **Subtract your result from $\bar{y}$.**

TIP

To save a great deal of time calculating the best-fitting line, keep in mind that five well-known summary statistics are all you need to do all the necessary calculations. I call them the "big-five statistics" (not to be confused with the five-number summary from Chapter 2):

1. The mean of the x-values (denoted $\bar{x}$)

2. The mean of the y-values (denoted $\bar{y}$)

3. The standard deviation of the x-values (denoted s_x)

4. The standard deviation of the y-values (denoted s_y)

5. The correlation between X and Y (denoted r)

(This chapter and Chapter 2 contain formulas and step-by-step instructions for these statistics.)

Interpreting the slope and y-intercept

Even more important than being able to calculate the slope and y-intercept to form the best-fitting regression line is the ability to interpret their values.

Interpreting the slope

The slope is interpreted in algebra as "rise over run." If the slope for example is 2, you can write this as 2/1 and say as X increases by 1, Y increases by 2, and that's how you move along from point to point on the line. In a regression context, the slope is the heart and soul of the equation because it tells you how much you can expect Y to change as X increases.

In general, the units for slope are the units of the Y-variable per units of the X-variable. It's a ratio of change in Y per change in X. Suppose in studying the effect of dosage level in milligrams (mg) on blood pressure, a researcher finds that the slope of the regression line is –2.5. You can write this as –2.5/1 and say blood pressure is expected to decrease by 2.5 points on average per 1 mg increase in drug dosage.

TIP

Always remember to use proper units when interpreting slope.

If using a 1 in the denominator of slope is not super-meaningful, you can multiply the top and bottom by any number (as long as it's the same number) and interpret it that way instead. In the blood pressure example, instead of writing slope as –2.5/1 and interpreting it as a decrease of 2.5 points per 1 mg increase of the drug, you can multiply the top and bottom by 10 to get –25/10 and say an increase in dosage of 10 mg results in a 25-point decrease in blood pressure.

Interpreting the y-intercept

The y-intercept is the place where the regression line $y = mx + b$ crosses the y-axis and is denoted by b (see earlier section "Finding the y-intercept"). Sometimes the y-intercept can be interpreted in a meaningful way, and sometimes not. This differs from slope, which is always interpretable. In fact, between the two elements of slope and intercept, the slope is the star of the show, with the y-intercept serving as the less famous but still noticeable sidekick.

There are times when the y-intercept makes no sense. For example, suppose you use rain to predict bushels per acre of corn; if the regression line crosses the y-axis somewhere below zero (and it most likely will), the y-intercept will make no sense. You can't have negative corn production.

Another situation when it's not okay to interpret the y-intercept is if there are no data near the point where $x = 0$. For example, suppose you want to use students' scores on Midterm 1 to predict their scores on Midterm 2. The y-intercept represents a prediction for Midterm 2 when the score on Midterm 1 is zero. You don't expect scores on a midterm to be at or near zero unless someone did not take the exam, in which case his or her score would not be included in the first place.

Many times, however, the y-intercept is of interest to you, it has meaning, and you have data collected in that area (where $x = 0$). For example, if you're predicting coffee sales at Green Bay Packer games using temperature, some games have temperatures at or even below zero, so predicting coffee sales at these temperatures makes sense. (As you might guess, they sell more and more coffee as the temperature dips.)

The best-fitting line for the crickets

The "big-five" statistics from the subset of cricket data are shown in Table 10-2.

TABLE 10-2 **Big-Five Statistics for the Cricket Data**

Variable	Mean	Standard Deviation	Correlation
# Chirps (x)	$\bar{x} = 26.5$	$s_x = 7.4$	$r = +0.98$
Temp (y)	$\bar{y} = 67$	$s_y = 6.8$	

The slope, m, for the best-fitting line for the subset of cricket chirp versus temperature data is $m = r\left(\dfrac{s_y}{s_x}\right) = 0.98\left(\dfrac{6.8}{7.4}\right) = 0.90$. So, as the number of chirps increases by 1 chirp per 15 seconds, the temperature is expected to increase by 0.90 degrees Fahrenheit on average. To get a more practical interpretation, you can multiply the top and bottom of the slope by 10 to get 9.0/10 and say that as chirps increase by 10 (per 15 seconds), temperature increases 9 degrees Fahrenheit.

Now, to find the y-intercept, b, you take $\bar{y} - m * \bar{x}$, or $67 - (0.90) * (26.5) = 43.15$. So the best-fitting line for predicting temperature from cricket chirps based on the data is $y = 0.90x + 43.15$, or temperature (in degrees Fahrenheit) 0.90 * (number of chirps in 15 seconds) +43.15. The y-intercept would try to predict temperature when there is no chirping going on at all. However, no data were collected at or near this point, so you can't make predictions for temperature in this area. You can't predict temperature using crickets if the crickets are silent.

Making Predictions

After you have a strong linear relationship, and you find the equation of the best-fitting line $y = mx + b$, you use that line to predict y for a given x-value. This amounts to plugging the x-value into the equation and solving for y. For example, if your equation is $y = 2x + 1$, and you want to predict y for $x = 1$, then plug 1 into the equation for x to get $y = 2(1) + 1 = 3$.

Remember that you choose the values of X (the explanatory variable) that you plug in; what you predict is Y, the response variable, which totally depends on X. By doing this, you are using one variable that you can easily collect data on, to predict a Y variable that is difficult or not possible to measure; this works well as long as X and Y are correlated. That's the big idea of regression.

From the previous section, the best-fitting line for the crickets is $y = 0.90x + 43.15$. Say you're camping, listening to crickets, and you remember that you can predict temperature by counting chirps. You count 35 chirps in 15 seconds. You put in 35 for x and find $y = 0.90(35) + 43.15 = 74.65$ degrees F. (Yeah, you memorized the formula just in case you needed it.) So, because crickets chirped 35 times in 15 seconds, you figure the temperature is probably about 75 degrees Fahrenheit.

Avoid Extrapolation!

Just because you have a model doesn't mean you can plug in any value for X and do a good job of predicting Y. For example, in the chirping data, there are no data collected for less than 18 chirps or more than 39 chirps per 15 seconds (refer back to Table 10-1). If you try to make predictions outside this range you're going into uncharted territory; the farther outside this range you go with your x-values, the more dubious your predictions for y will get. Who's to say the line still works outside of the area where data were collected? Do you think crickets will chirp faster and faster without limit? At some point, they would either pass out or burn up!

Making predictions using x-values that fall outside the range of your data is a no-no. Statisticians call this *extrapolation*; watch for researchers who try to make claims beyond the range of their data.

Correlation Doesn't Necessarily Mean Cause-and-Effect

Scatterplots and correlations identify and quantify relationships between two variables. However, if a scatterplot shows a definite pattern, and the data are found to have a strong correlation, that doesn't necessarily mean that a cause-and-effect relationship exists between the two variables. A *cause-and-effect relationship* is one where a change in X causes a change in Y. (In other words, the change in Y is not only associated with a change in X, it is directly caused by X.)

For example, suppose a well-controlled medical experiment is conducted to determine the effects of dosage of a certain drug on blood pressure. (See a total breakdown of experiments in Chapter 13.) The researchers look at their scatterplot and see a definite downhill linear pattern; they calculate the correlation and it's strong. They conclude that increasing the dosage of this drug causes a decrease in blood pressure. This cause-and-effect conclusion is okay because they controlled for other variables that could affect blood pressure in their experiment, such as other drugs taken, age, general health, and so on.

However, if you made a scatterplot and examined the correlation between ice cream consumption versus murder rates, you would also see a strong linear relationship (this time uphill). Yet no one would claim that more ice cream consumption causes more murders to occur.

What's going on here? In the drug example, the data were collected through a well-controlled medical experiment, which minimizes the influence of other factors that might affect blood pressure changes. In the second example, the data were just based on observation, and no other factors were examined. It turns out that this strong relationship exists because increases in murder rates and ice cream sales are both related to increases in temperature.

(Temperature in this case is called a *confounding variable*; it affects both X and Y but was not included in the study — see Chapter 13.)

TECHNICAL STUFF

Whether two variables are found to be causally associated depends on how the study was conducted. Only a well-designed experiment (see Chapter 13) or a large collection of several different observational studies can show enough evidence for cause-and-effect.

Yet, this condition is often ignored as the media gives us headlines such as "Doctors can lower malpractice lawsuits by spending more time with patients." In reality, it was found that doctors who have fewer lawsuits are the type of doctor who spends a lot of time with patients. But that doesn't mean taking a bad doctor and having him spend more time with his patients will reduce his malpractice suits; in fact, spending more time with him might create even more problems.

And you can't say that crickets chirping faster will cause the temperature to increase, of course, but you do know you can count cricket chirps and do a pretty good job predicting temperature nonetheless, through simple linear regression.

IN THIS CHAPTER

» Organizing probabilities in two-way
tables

» Figuring marginal, conditional, and joint
probabilities

» Checking for independence

Chapter **11**
Two-Way Tables

C ategorical variables place individuals into groups based on
certain possible outcomes. For example, gender (male,
female) whether you ate breakfast this morning (yes, no),
or political affiliation (Democrat, Republican, Independent,
Other). Oftentimes you look for relationships between two categorical variables; for example, "Are females more likely to eat
breakfast than males?" A two-way table classifies individuals into
groups based on all possible pairs of outcomes of two categorical
variables (for example, male breakfast eaters, female breakfast
eaters, and so on). In this chapter, you see how two-way tables
help you organize and figure probabilities and check for independence of two events.

Organizing and Interpreting
a Two-Way Table

Suppose you are a basketball nut and you love to watch your favor-
ite player shoot free throws. After watching him shoot pairs of free
throws for a long time, you notice two things. First, it seems like
he makes the second shot more often than he makes the first. You
also believe, based on your observations, that when he misses the

first shot, he makes the second one even more often. You always thought that free throw attempts were independent and that the outcome of one shot didn't influence the outcome of another, but in this case, you suspect there is a relationship after all, for this player at least. So you launch your own statistical investigation to find out.

Suppose you collect data on this player during 155 different trips to the free throw line. Each time he shoots a pair of free throws you record the outcomes. Examining your data, you see he made the first shot and missed the second one 40 times; 60 times he made both free throws; 10 times he missed both; and 45 times he missed the first one and made the second.

The next step is to organize your data into a two-way table. The following sections take you through it.

Defining the outcomes

The first step in setting up a two-way table is to define the sample space and the outcomes of the experiment using probability notation. In the free throw example, your first categorical variable is the outcome of the first throw. This variable has two possible outcomes: 1) he made the first free throw (indicated by Y_1); or 2) he missed the first free throw (indicated by N_1). Similarly, the second categorical variable is the outcome of the second shot; its outcomes, Y_2 and N_2, represent making and missing the second shot, respectively.

The sample space, S, lists all possible pairs of outcomes of this two-variable data. Because each variable has 2 possible outcomes, there are $2*2 = 4$ pairs of possible outcomes for the pair of free throws:

$$S = \{Y_1Y_2; Y_1N_1; N_2; N_1Y_2; \text{and } N_1N_2\}$$

Setting up the rows and columns

You can organize the two-way table using rows to represent one variable (the outcome of the first free throw) and columns to represent the other variable (the outcome of the second free throw). Table 11-1 shows what the two-way table looks like.

TABLE 11-1 **Two-Way Table Set-up for Pairs of Free Throws**

	Made Second Free Throw (Y_2)	Missed Second Free Throw (N_2)
Made First Free Throw (Y_1)	$Y_1 \cap Y_2$	$Y_1 \cap N_2$
Missed First Free Throw (N_1)	$N_1 \cap Y_2$	$N_1 \cap N_2$

Notice the table has $2 * 2 = 4$ boxes in it. These boxes are called the cells of the two-way table. Each *cell* represents an intersection of a row and column. For example the cell in the upper right-hand corner of the table represents the outcome where the player made the first free throw and missed the second one. In probability notation, this represents the intersection of the outcomes Y_1 and N_2, written as $Y_1 \cap N_2$. I wrote in the events represented by each cell of the free throw two-way table in Table 11-1.

Inserting the numbers

Remember that the player made the first free throw and missed the second a total of 40 times; 60 times he made both free throws; 10 times he missed both; and 45 times he missed the first one and made the second. Now enter the basketball data into a two-way table and calculate probabilities.

Looking at the labels on the rows and columns, you see 60 goes into the upper left cell (represented by the event $Y_1 \cap Y_2$), 40 goes into the upper right cell (represented by the event $Y_1 \cap N_2$), 45 is in the bottom left cell (represented by $N_1 \cap Y_2$), and 10 is in the bottom right (represented by the event $N_1 \cap N_2$). The number of individuals inside of a cell in row i and column j of a two-way table is called the *cell count* for the $(i, j)^{\text{th}}$ cell. Table 11-2 shows the two-way table with the cell counts.

Finding the row, column, and grand totals

Once the cell counts are placed within a two-way table, it's always a good idea to total the rows and columns and write those totals in the margins. These are aptly named *marginal totals*. You can see in Table 11-2 that the total for the first row, $60 + 40 = 100$, is written in the "Row Totals" column of the first row. This means the total number of times the first shot was successful was 100 (no matter what happened on the second shot). Similarly, the row total for row 2 represents the total number of times the first shot was missed regardless of what happened on the second shot $(45 + 10 = 55)$.

The column totals are also included, listed in the row at the bottom of Table 11-2. The first column total is $60 + 45 = 105$, which represents the total number of times a basket was made on the second shot (whether or not a basket was made on the first shot.) The second column total represents the total number of times the second shot was missed (regardless of what happened on the first shot), $40 + 10 = 50$. Notice that the row totals sum to the grand total of 155, the total number of pairs of shots attempted. Similarly the column totals sum to the grand total of 155. The row, column, and grand totals are all shown in Table 11-2.

Finding Probabilities within a Two-Way Table

Once the two-way table is set up, you can begin using it to calculate probabilities and answer important questions about various events. For example, what is the probability that a player makes both free throws? That he makes the first one? That he makes the second free throw given that he misses the first one? And if he misses the first free throw, does that affect his chances of making the second one?

Figuring joint probabilities

A *joint probability* is the probability of the intersection of two outcomes or events. For example, the probability that a player makes the first free throw *and* the second free throw is a joint probability and is denoted $P(Y_1 \cap Y_2)$. The word *and* provides a clue that it's a joint probability, as in "*A and B.*"

Finding joint probabilities is easy using a two-way table, because the cells of the two-way table already show the number of individuals in each intersection. To find the probability of any intersection, take the number in that cell and divide by the grand total (found in the lower right corner of the two-way table). For example, the probability that a player makes the first free throw and the second, $P(Y_1 \cap Y_2)$ is the number in the upper left cell of the

TABLE 11-2 Two-Way Table for of Cell Counts for Pairs of Free Throws

	Made Second Free Throw (Y_2)	Missed Second Free Throw (N_2)	Row Totals
Made First Free Throw (Y_1)	60	40	$60 + 40 = 100$
Missed First Free Throw (N_1)	45	10	$45 + 10 = 55$
Column Totals	$60 + 45 = 105$	$40 + 10 = 50$	Grand Total = 155

two-way table, 60, divided by the grand total, 155. That means the probability of making both free throws is $60 / 155 = 0.39$, or 39%.

The general formula for finding a joint probability using a two-way table is $\dfrac{\text{number in cell}(i, j)}{\text{grand total}}$, where the cell in the jth row and the jth column is denoted by cell (i, j).

Calculating marginal probabilities

A marginal probability is the probability of one outcome or event occurring, regardless of what happened with any other variables. For example, the probability that a player makes the first free throw (regardless of what happens on the second shot) is the marginal probability of Y_1 and is denoted $P(Y_1)$. And the probability that a player makes the second free throw (regardless of what happened on the first shot) is the marginal probability of Y_2, denoted $P(Y_2)$.

To find the marginal probability of any single event, take the number in the corresponding row or column total and divide by the grand total. For example, look at the event that the player makes the first free throw (regardless of what happens on the second shot). This event is represented by row 1 of the two-way table, denoted Y_1 (see Table 11-2). So the probability that a player makes the first free throw, $P(Y_1)$, is found by taking the row 1 total, 100, and dividing by the grand total, 155, to get 0.65 or 65%.

Now look at the event that the player makes the second free throw. (Notice there is no mention of what happens on the first shot, so that tells you it's a marginal probability.) This event is represented in column 1 of the two-way table, denoted Y_2 (see Table 11-2). So the probability that a player makes the second free throw, $P(Y_2)$, is found by taking the column 1 total, 105, and dividing by the grand total, 155, to get $105 / 155 = 0.68$ or 68%. So your first observation is true; he makes the second free throw more often than the first one (68% compared to 65%).

The general formula for finding the marginal probability of an event in row i of the two-way table is $P(\text{row } i \text{ event}) = \dfrac{\text{row } i \text{ total}}{\text{grand total}}$. The general formula for finding the marginal probability of an event in column j of the two-way table is $P(\text{column } j \text{ event}) = \dfrac{\text{column } j \text{ total}}{\text{grand total}}$.

Finding conditional probabilities

A *conditional probability* is the probability of one event happening, given that the outcome of another event is known. For example, the probability that a player makes the second free throw, given that he made the first one, is the conditional probability of Y_2 given Y_1 and is denoted $P(Y_2 \mid Y_1)$. And the probability that a player misses the second free throw given that he missed the first one is the conditional probability of N_2 given N_1 and is denoted $P(N_2 \mid N_1)$.

The formula for the conditional probability of A given B is $P(A \mid B) = \dfrac{P(A \cap B)}{P(B)}$. You're dividing by $P(B)$ because you know event B has happened, making this the new sample space. But because the denominators of $P(A \cap B)$ and $P(B)$ both equal the grand total in the two-way table, you can find the conditional probability by just taking the number in the cell representing $A \cap B$, divided by the appropriate row or column total for event B. (The denominators for these probabilities are the same, the grand total, and they cancel out when you divide the probabilities, so you don't have to include them in the calculations.)

For example, look at the event that the player makes the second free throw given he made the first one. This event is denoted $Y_2 \mid Y_1$. (Y_1 is the event that is known, so you use its marginal total in the denominator.) So the probability that a player makes the second free throw given he made the first one, $P(Y_2 \mid Y_1)$, is found by taking the number in the cell representing $Y_2 \cap Y_1$, 60, and divide by the row total representing Y_1, 100. So the probability of making the second shot, given he made the first, is 0.60 or 60%.

Now look at the event that the player misses the second free throw given he missed the first one. This event is denoted $N_2 \mid N_1$. So the probability that a player misses the second free throw given he missed the first one, $P(N_2 \mid N_1)$, is found by taking the number in the cell for $N_2 \cap N_1$, 10, and dividing by the second row total, 55, to get 0.18 or 18%.

The general formula for finding the conditional probability of an event in row i given an event in row j of the two-way table is $P(\text{row } i \mid \text{column } j) = \dfrac{\text{number in cell}(i, j)}{\text{column } j \text{ total}}$. The general formula for finding the conditional probability of the event in column j given the event in row i of the two-way table is $P(\text{column } j \mid \text{row } i) = \dfrac{\text{number in cell}(i, j)}{\text{column } i \text{ total}}$.

TIP

A conditional probability is the probability of one event happening given another event is known to have occurred. Using a two-way table, to find the conditional probability of an event in a certain column of the table given an event in a row of the table, take the cell count for the intersection divided by the corresponding row total. To find the conditional probability of an event in a certain row of the table given an event in a column of the table, take the cell count for the intersection divided by the corresponding column total. The clue that it's a conditional probability is the fact that you know one event is known to have occurred. Words like *given, knowing,* and *of* are often used to mean conditional probability.

Here is a summary of the conditional probabilities you can calculate regarding the free throws example:

>> The probability of making the second shot given he made the first one, $P(Y_2 \mid Y_1)$, is $60 / 100 = 0.60$ or 60%.

>> The probability of missing the second shot given he made the first one, $P(N_2 \mid Y_1)$, is $40 / 100 = 0.40$ or 40%. Observe that this is $1 - 0.60$, because you either miss the second or you don't. (That is, making the shot and missing the shot are complements of each other.)

>> The probability of missing the second shot given he missed the first one, $P(N_2 \mid N_2)$, is $10 / 55 = 0.18$ or 18%.

>> The probability of making the second shot given he missed the first one, $P(Y_2 \mid N_1)$, is $45 / 55 = 0.82$ or 82%. Note this is the complement of the previous event, so their probabilities sum to one.

Once you're given that event B has happened, A either happens or it doesn't. So it is true that $P(A \mid B) + P(A^c \mid B) = 1$. But it is *not* true that $P(A \mid B) + P(A \mid B^c) = 1$ because in each term you are conditioning on a different event. This is a common mistake that you definitely want to avoid. (The notation statisticians use for the event where A doesn't happen is A^c. We call it "A complement.")

Checking for Independence

You know he makes the second free throw more often than the first, from the section on marginal probabilities. Now you are ready to answer your second question: In situations where he misses the first shot, does he make the second shot even more often? If the answer is yes, then we say the outcome of the second shot is related to, or is *dependent*, on, the outcome of the first shot. If the answer is no, then we say the outcome of the second shot is not related to, or is *independent* of, the outcome of the first.

Formally speaking, two events A and B are *independent* if $P(A \mid B) = P(A)$. In other words, if the knowledge that B has happened does not change the probability of A happening, then events A and B are independent. Note this also means that if events A and B are independent, then $P(A \mid B) = P(A \mid B^c)$, because both of these terms must be equal to $P(A)$ in that case.

To summarize these results, you can show events A and B are independent using two different methods:

>> Method 1: If $P(A \mid B) = P(A)$ then A and B are *independent*. If they are not equal, then we say A and B are *dependent*.

>> Method 2: If $P(A \mid B) = P(A \mid B^c)$ then A and B are *independent*. If they are not equal, then we say A and B are *dependent*.

You only have to check for independence using one of those methods; you need not use both.

Using method 1 to answer your question, you check for independence by comparing his overall rate of making the second shot to his rate of making the second shot when you know he's missed the first. That is, check to see if $P(Y_2) = P(Y_2 \mid N_1)$. You know that $P(Y_2)$ is the overall chance of making the second shot, which equals $150 / 155 = 0.68\%$. Now if the first shot was missed, the probability of making the second shot increases to $P(Y_2 \mid N_1) = 45 / 55 = 0.82$ or 82%. Because 0.68 is not equal to 0.82, the outcomes of the two shots are dependent. In situations where the first free throw is missed, he makes the second one more often than his overall rate.

Using method 2, you check to see if the probability of making the second shot is the same whether the first shot is made or missed. That is, check to see if $P(Y_2 \mid Y_1) = P(Y_2 \mid N_1)$. When the first shot is made, the chance of making the second is $P(Y_2 \mid Y_1) = 60 / 100 = 0.60$ or 60%; when the first shot is missed, the chance of making the second shot increases to $P(Y_2 \mid N_1) = 45 / 55 = 0.82$ or 82%. Since these probabilities are not equal, the outcomes of the two shots are dependent. The probability of making the second shot is higher when he misses the first one than when he makes the first one.

TECHNICAL
STUFF

Although the pairs of probabilities being compared are different for the two methods of checking for independence, the overall conclusions should always agree. In this example, your suspicions were right — no matter how you slice it, your player is more likely to make the second shot when he misses the first than when he makes the first.

Chapter **12**

A Checklist for Samples and Surveys

Surveys are all around you — I guarantee that at some point in your life, you'll be asked to complete a survey. You're also likely to be inundated with the results of surveys, and before you consume their information, you need to evaluate whether they were properly designed. In this chapter, I present a checklist you can use to evaluate or plan a survey.

The survey process can be broken down into a series of ten elements that should be checked:

1. Target population is well defined.
2. Sample matches the target population.
3. Sample is randomly selected.
4. Sample size is large enough.
5. Nonresponse is minimized.
6. Type of survey is appropriate.
7. Questions are well worded.
8. Survey is properly timed.
9. Personnel are well trained.
10. Proper conclusions are made.

This list helps you carry out your own survey or critique someone else's survey. In the following sections, I address each item and discuss its role in getting a good survey done.

The Target Population Is Well Defined

The *target population* is the entire group of individuals that you're interested in studying. For example, suppose you want to know what the people in Great Britain think of reality TV. The target population is all the residents of Great Britain.

Many researchers don't do a good job of defining their target populations clearly. For example, if the American Egg Board wants to say "Eggs are good for you!" it needs to specify who the "you" is. For example, is the Egg Board prepared to say that eggs are good for people who have high cholesterol? What if one of the studies the group cites is based only on young people who are healthy and eating low-fat diets — is that who they mean by "you"?

TECHNICAL STUFF

If the target population isn't well defined, the survey results are likely to be biased. The sample that's actually studied may contain people outside the intended population, or the survey may exclude people who should have been included.

The Sample Matches the Target Population

When you're conducting a survey, you typically can't ask every single member of the target population to provide the information you're looking for. The best you can do is select a good *sample* (a subset of individuals from the population) and get the information from them. A good sample *represents* the target population. The sample doesn't systematically favor certain groups within the target population, and it doesn't systematically exclude certain people, either.

TIP

The best scenario for selecting a representative sample is to obtain a *sampling frame* — a list of all the members of the target population — and draw randomly from that. If such a list isn't possible, you need some mechanism that gives everyone in the population an equal opportunity to be chosen to participate in the survey. For example, if a house-to-house survey of a city is needed, an updated map including all houses in that city should be used as the sampling frame.

The Sample Is Randomly Selected

An important feature of a good study is that the sample is randomly selected from the target population. Randomly means that every member of the target population has an equal chance of being included in the sample. In other words, the process you use for selecting your sample can't be biased.

The biggest problem to watch for is convenience samples. A *convenience sample* is a sample selected in a way that's easiest on the researcher — for example call-in polls, man-on-the-street surveys, or Internet surveys. Convenience samples are totally nonrandom, and their results are not credible.

For surveys involving people, reputable polling organizations such as the Gallup Organization use a random digit dialing procedure to telephone the members of their sample. This excludes people without phones, of course, so this kind of survey does have a bit of bias. In this case, though, most people do have phones (over 95%, according to the Gallup Organization), so the bias against people who don't have phones is not a big problem.

The Sample Size Is Large Enough

You've heard the saying, "Less is more"? With surveys, the saying is, "Less good information is better than more bad information, but more good information is better."

If you have a large sample size, and the sample is representative of the target population (meaning randomly selected), you can count on that information to be pretty accurate. Exactly how accurate depends on the sample size, but in general a bigger sample leads to more accurate information (assuming the data are well collected).

TIP

A quick and dirty formula to calculate the accuracy of a survey is to divide by the square root of the sample size. For example, a survey of 1,000 (randomly selected) people is accurate to within $\pm \frac{1}{\sqrt{1,000}}$, which is 0.032 or 3.2%. This percentage is called the *margin of error*. (Note that this formula is just a *rough* estimate. A better estimate can be found using the formulas from Chapter 7.)

TECHNICAL STUFF

Beware of surveys that have a large sample size but it's not randomly selected; Internet surveys are the biggest culprit. A company can say that 50,000 people logged on to its website to answer a survey, but that information is biased, because it represents opinions of those who had access to the Internet, went to the website, and chose to complete the survey.

Nonresponse Is Minimized

After the sample size has been chosen and the sample of individuals has been randomly selected from the target population, you have to get the information you need from the people in the sample. If you've ever thrown away a survey or refused to answer a few questions over the phone, you know that getting people to participate in a survey isn't easy.

The importance of following up

If a researcher wants to minimizes bias, the best way to handle nonresponse is to "hound" the people in the sample: Follow up one, two, or even three times, offering dollar bills, coupons, self-addressed stamped return envelopes, chances to win prizes, and so on. Note that offering more than a small token of incentive and appreciation for participating can create bias as well, because then people who really need the money are more likely to respond than those who don't.

Consider what motivates *you* to fill out a survey. If the incentive provided by the researcher doesn't get you, maybe the subject matter piques your interest. Unfortunately, this is where bias comes in. If only those folks who feel very strongly respond to a survey, only their opinions will count; because the other people who don't really care about the issue don't respond, each "I don't care" vote doesn't count. And when people do care but don't take the time to complete the survey, those votes don't count, either.

The response rate of a survey is a percentage found by taking the number of respondents divided by the total sample size and multiplying by 100%. The ideal response rate according to statisticians is anything over 70%. However, most response rates fall well short of that, unless the survey is done by a very reputable organization, such as Gallup.

Look for the response rate when examining survey results. If the response rate is too low (much less than 70%), the results may be biased and should be ignored. Selecting a smaller initial sample and following up aggressively is better than selecting a bigger sample that ends up with a low response rate. Plan several follow-up calls/mailings to reduce bias. It also helps increase the response rate to let people know up front whether their results will be shared or not.

Anonymity versus confidentiality

If you were to conduct a survey to determine the extent of personal email usage at work, the response rate would probably be low because many people are reluctant to disclose their use of personal email in the workplace. You could encourage people to respond by letting them know that their privacy would be protected during and after the survey.

When you report the results of a survey, you generally don't tie the information collected to the names of the respondents, because doing so would violate the privacy of the respondents. You've probably heard the terms *anonymous* and *confidential* before, but you may not realize that they have totally different meanings in terms of privacy issues. Keeping results confidential means that I could tie your information to your name in my report, but

I promise that I won't do that. Keeping results *anonymous* means that I have no way of tying your information to your name in my report, even if I wanted to.

If you're asked to participate in a survey, be sure you're clear about what the researchers plan to do with your responses and whether or not your name can be tied to the survey. (Good surveys always make this issue very clear for you.) Then make a decision as to whether you still want to participate.

The Survey Is of the Right Type

Surveys come in many types: mail surveys, telephone surveys, Internet surveys, house-to-house interviews, and man-on-the-street surveys (in which someone comes up to you with a clipboard and asks, "Do you have a few minutes to participate in a survey?"). One very important yet sometimes overlooked criterion of a good survey is whether the type of survey being used is appropriate for the situation. For example, if the target population is the population of people who are visually impaired, sending them a survey in the mail that has a tiny font isn't a good idea (yes, this has happened!).

When looking at the results of a survey, be sure to find out what type of survey was used and reflect on whether this type of survey was appropriate.

Questions Are Well Worded

The way in which a question is worded in a survey can affect the results. For example, while President Bill Clinton was in office and the Monica Lewinsky scandal broke, a CNN/Gallup Poll conducted August 21–23, 1998, asked respondents to judge Clinton's favorability, and about 60% gave him a positive result. When CNN/Gallup reworded the question to ask respondents to judge Clinton's favorability "as a person," only about 40% gave him a positive rating. These questions were both getting at the same issue; even though they were worded only slightly differently you can see how different the results are. So question wording does matter.

One huge problem is the use of misleading questions (in other words, questions that are worded in such a way that you know how the researcher wants you to answer). An example of a misleading question is, "Do you agree that the president should have the power of a line-item veto to eliminate waste?" This question should be worded in a neutral way, such as "What is your opinion about the line-item veto ability of a president?" Then give a scale from 1 to 5 where 1 = strongly disagree and 5 = strongly agree.

When you see the results of a survey that's important to you, ask for a copy of the questions that were asked and analyze them to ensure that they were neutral and minimized bias.

The Timing Is Appropriate

The timing of a survey is everything. Current events shape people's opinions, and while some pollsters try to determine how people really feel, others take advantage of these situations, especially the negative ones. For example, polls regarding gun control often come out right after a shooting that is reported by the national media. Timing of any survey, regardless of the subject matter, can still cause bias. Check the date when a survey was conducted and see whether you can determine any relevant events that may have temporarily influenced the results.

Personnel Are Well Trained

The people who actually carry out surveys have tough jobs. They have to deal with hang-ups, take-us-off-your-list responses, and answering machines. After they do get a live respondent at the other end of the phone or face to face, the job becomes even harder. For example, if the respondent doesn't understand the question and needs more information, how much can you say, while still remaining neutral?

For a survey to be successful, the survey personnel must be trained to collect data in an accurate and unbiased way. The key is to be clear and consistent about every possible scenario that may come up, discuss how they should be handled, and have this discussion well before participants are ever contacted.

TIP

You can also avoid problems by running a pilot study (a practice run with only a few respondents) to make sure the survey is clear and consistent and that the personnel are handling responses appropriately. Any problems identified can be fixed before the real survey starts.

Proper Conclusions Are Made

Even if a survey is done correctly, researchers can misinterpret or over-interpret results so that they say more than they really should. Here are some of the most common errors made in drawing conclusions from surveys:

>> Making projections to a larger population than the study actually represents

>> Claiming a difference exists between two groups when a difference isn't really there

>> Saying that "these results aren't scientific, but . . ." and then presenting the results as if they are scientific

To avoid common errors made when drawing conclusions:

1. **Check whether the sample was selected properly and that the conclusions don't go beyond the population presented by that sample.**

2. **Look for disclaimers about surveys *before* reading the results, if you can.**

   That way, you'll be less likely to be influenced by the results if, in fact, the results aren't based on a scientific survey. Now that you know what a *scientific survey* (the media's term for an accurate and unbiased survey) actually involves, you can use those criteria to judge whether survey results are credible.

3. **Be on the lookout for statistically incorrect conclusions.**

   If someone reports a difference between two groups based on survey results, be sure the difference is larger than the reported margin of error. If the difference is within the margin of error, you should expect the sample results to vary by that much just by chance, and the so-called "difference" can't really be generalized to the entire population; see Chapter 7.

4. **Tune out anyone who says, "These results aren't scientific, but. . . ."**

Know the limitations of any survey and be wary of any information coming from surveys in which those limitations aren't respected. A bad survey is cheap and easy to do, but you get what you pay for. Before looking at the results of any survey, investigate how it was designed and conducted, so that you can judge the quality of the results.

Chapter **13**

A Checklist for Judging Experiments

I n this chapter, you go behind the scenes of experiments — the driving force of medical studies and other investigations in which comparisons are made. You find out the difference between experiments and observational studies and discover what experiments can do for you, how they're supposed to be done, and how you can spot misleading results.

Experiments versus Observational Studies

Although many different types of studies exist, you can boil them all down to basically two different types: experiments and observational studies. An *observational study* is just what it sounds like: a study in which the researcher merely observes the subjects and records the information. No intervention takes place, no changes are introduced, and no restrictions or controls are imposed. For example, a survey is an observational study. An *experiment* is a study that doesn't simply observe subjects in their natural state, but deliberately applies treatments to them in a controlled situation and records the outcomes (for example, medical studies done in a laboratory). Experiments are generally more powerful than observational studies; for example, an experiment can identify a cause-and-effect relationship between two variables, whereas an observational study can only point out a connection.

Criteria for a Good Experiment

To decide whether an experiment is credible, check the following items:

1. **Is the sample size large enough to yield precise results?**
2. **Do the subjects accurately represent the intended population?**
3. **Are the subjects randomly assigned to the treatment and control groups?**
4. **Was the placebo effect measured (if applicable)?**
5. **Are possible confounding variables controlled for?**
6. **Is the potential for bias minimized?**
7. **Were the data analyzed correctly?**
8. **Are the conclusions appropriate?**

In the following sections, I present action items for evaluating an experiment based on each of the above criteria.

Inspect the Sample Size

The size of a sample greatly affects the accuracy of the results. The larger the sample size, the more accurate the results are, and the more powerful the statistical analysis will be at detecting real differences due to treatments.

Small samples — small conclusions

You may be surprised at the number of research headlines that were based on very small samples. If the results are important to you, ask for a copy of the research report and find out how many subjects were involved in the study.

Also be wary of research that finds significant results based on very small sample sizes (especially those much smaller than 30). It could be a sign of what statisticians call *data fishing*, where someone fishes around in his data set using many different kinds of analyses until he finds a significant result (which is not repeatable because it was just a fluke).

Original versus final sample size

Be specific about what a researcher means by *sample size*. For example, ask how many subjects were selected to participate in an experiment and then ask for the number who actually completed the experiment — these two numbers can be very different. Make sure the researchers can explain any situations in which the research subjects decided to drop out or were unable (for some reason) to finish the experiment.

An article in the *New York Times* entitled "Marijuana Is Called an Effective Relief in Cancer Therapy" says in the opening paragraph that marijuana is "far more effective" than any other drug in relieving the side effects of chemotherapy. When you get into the details, you find out that the results are based on only 29 patients (15 on the treatment, 14 on a placebo). To add to the confusion, you find out that only 12 of the 15 patients in the treatment group actually completed the study; so what happened to the other 3 subjects?

Examine the Subjects

An important step in designing an experiment is selecting the sample of participants, called the research *subjects*. Although researchers would like for their subjects to be selected randomly from their respective populations, in most cases this just isn't possible. For example, suppose a group of eye researchers wants to test out a new laser surgery on nearsighted people. To select their subjects, they randomly select various eye doctors from across the country and randomly select nearsighted patients from these doctors' files. They call up each person selected and say, "We're experimenting with a new laser surgery treatment for nearsightedness, and you've been selected at random to participate in our study. When can you come in for the surgery?" This may sound like a good random sampling plan, but it doesn't make for an ethical experiment.

The point is, getting a truly random sample of people to participate in an experiment would be great, but is typically not feasible or ethical to do. Rather than select people at random, experimenters do the best they can to gather volunteers who meet certain criteria so they're doing the experiment on an appropriate cross-section of the population. The randomness part comes in when individuals are assigned to the groups (treatment group, control group, and so forth) in a random fashion, as explained in the next section.

Check for Random Assignments

After the sample has been selected, the subjects are assigned to either a *treatment group,* which receives a certain level of some factor being studied, or a *control group,* which receives either no treatment or a fake treatment. How the subjects are assigned to their respective groups is extremely important.

Suppose a researcher wants to determine the effects of exercise on heart rate. The subjects in his treatment group run five miles and have their heart rates measured before and after the run. The subjects in his control group will sit on the couch the whole time and watch reruns of *The Simpsons.* If only the health nuts (who probably already have excellent heart rates) volunteer to be in the treatment group, the researcher will be looking only at the effect of the treatment (running five miles) on very healthy and active people. He won't see the effect that running five miles has on the heart rates of couch potatoes. This nonrandom assignment of subjects to the treatment and control groups can have a huge impact on his conclusions.

To avoid bias, subjects must be assigned to treatment/control groups at random. This results in groups that are more likely to be fair and balanced, yielding more credible results.

Gauge the Placebo Effect

A fake treatment takes into account what researchers call the placebo effect. The *placebo effect* is a response that people have (or think they're having) because they know they're getting some sort of "treatment" (even if that treatment is a fake treatment, also known as a placebo, such as sugar pills).

If the control group is on a placebo, you may expect them not to report any side effects, but you would be wrong. Placebo groups often report side effects in percentages that seem quite high; this is because the knowledge that some treatment is being taken (even if it's a fake treatment) can have a psychological (even a physical) effect. If you want to be fair about examining the side effects of a treatment, you have to take into account the side effects that the control group reports; that is, side effects that are due to the placebo effect.

TECHNICAL STUFF

In some situations, such as when the subjects have very serious diseases, offering a fake treatment as an option may be unethical. When ethical reasons bar the use of fake treatments, the new treatment is compared to an existing or standard treatment that is known to be effective. After researchers have enough data to see that one of the treatments is working better than the other, they will generally stop the experiment and put everyone on the better treatment, again for ethical reasons.

Identify Confounding Variables

A *confounding variable* is a characteristic which was not included or controlled for in the study, but can influence the results. That is, the real effects due to the treatment are confounded, or clouded, due to this variable.

For example, if you select a group of people who take vitamin C daily, and a group who don't, and follow them all for a year's time counting how many colds they get, you might notice the group taking vitamin C had fewer colds than the group who didn't take vitamin C. However, you cannot conclude that vitamin C reduces colds. Because this was not a true experiment but rather an observational study, there are many confounding variables at work. One possible confounding variable is the person's level of health consciousness; people who take vitamins daily may also wash their hands more often, thereby heading off germs.

How do researchers handle confounding variables? Control is what it's all about. Here you could pair up people who have the same level of health-consciousness and randomly assign one person in each pair to taking vitamin C each day (the other person gets a fake pill). Any difference in number of colds found between the groups is more likely due to the vitamin C, compared to the original observational study. Good experiments control for potential confounding variables.

Assess Data Quality

To decide whether or not you're looking at credible data from an experiment, look for these characteristics:

>> **Reliability:** Reliable data get repeatable results with subsequent measurements. If your doctor checks your weight once and you get right back on the scale and see it's different, there is a reliability issue. Same with blood tests, blood pressure and temperature measurements, and the like. It's important to use well-calibrated measurement instruments in an experiment to help ensure reliable data.

>> **Unbiasedness:** Unbiased data contains no systematic favoritism of certain individuals or responses. Bias is caused in many ways: by a bad measurement instrument, like a bathroom scale that's sometimes 5 pounds over; a bad sample, like a drug study done on adults when the drug is actually taken by children; or by researchers who have preconceived expectations for the results ("You feel better now after you took that medicine, don't you?").

Bias is difficult, and in some cases even impossible, to measure. The best you can do is anticipate potential problems and design your experiment to minimize them. For example, a *double-blind* experiment means that neither the subjects nor the researchers know who got which treatment or who is in the control group. This is one way to minimize bias by people on either side.

>> **Validity:** Valid data measure what they are intended to measure. For example, reporting the prevalence of crime using number of crimes in an area is not valid; the *crime rate* (number of crimes per capita) should be used because it factors in how many people live in the area.

Check out the Analysis

After the data have been collected, they're put into that mysterious box called the *statistical analysis*. The choice of analysis is just as important (in terms of the quality of the results) as any other aspect of a study. A proper analysis should be planned in advance, during the design phase of the experiment. That way, after the data are collected, you won't run into any major problems during the analysis.

As part of this planning you have to make sure the analysis you choose will actually answer your question. For example, if you want to estimate the average blood pressure for the treatment group, use a confidence interval for one population mean (see Chapter 7). However, if you want to compare the average blood pressure for the treatment group versus a control group, you use a hypothesis test for two means (see Chapter 8). Each analysis has its own particular purpose; this book hits the highlights of the most commonly used analyses.

You also have to make sure that the data and your analysis are compatible. For example, if you want to compare a treatment group to a control group in terms of the amount of weight lost on a new (versus an existing) diet program, you need to collect data on how much weight each person lost (not just each person's weight at the end of the study).

Scrutinize the Conclusions

Some of the biggest statistical mistakes are made after the data have all been collected and analyzed — when it's time to draw conclusions, some researchers get it all wrong. The three most common errors in drawing conclusions are the following:

>> Overstating their results

>> Making connections or giving explanations that aren't backed up by the statistics

>> Going beyond the scope of the study in terms of whom the results apply to

Overstated results

When you read a headline or hear about the big results of the latest study, be sure to look further into the details of the study — the actual results might not be as grand as what you were led to believe. For example, suppose a researcher finds a new procedure that slows down tumor growth in lab rats. This is a great result but it doesn't mean this procedure will work on humans, or will be a cure for cancer. The results have to be placed into perspective.

Ad-hoc explanations

Be careful when you hear researchers explaining why their results came out a certain way. Some after-the-fact ("ad-hoc") explanations for research results are simply not backed up by the studies they came from. For example, suppose a study observes that people who drink more diet cola sleep fewer hours per night on average. Without a more in-depth study, you can't go back and explain why this occurs. Some researchers might conclude the caffeine is causing insomnia (okay. . .), but could it be that diet cola lovers (including yours truly) tend to be night owls, and night owls typically sleep fewer hours than average?

Generalizing beyond the scope

You can only make conclusions about the population that's represented by your sample. If you want to draw conclusions about the opinions of all Americans, you need a random sample of Americans. If your random sample came from a group of students in your psychology class, however, then the opinions of your psychology class is all you can draw conclusions about.

Some researchers try to draw conclusions about populations that have a broader scope than their sample, often because true representative samples are hard to get. Find out where the sample came from before you accept broad-based conclusions.

Chapter **14**

Ten Common Statistical Mistakes

This book is not only about understanding statistics that you come across in your job and everyday life; it's also about deciding whether the statistics are correct, reasonable, and fair. After all, if you don't critique the information and ask questions about it, who will? In this chapter, I outline some common statistical mistakes made out there, and I share ways to recognize and avoid those mistakes.

Misleading Graphs

Many graphs and charts contain misinformation, mislabeled information, or misleading information, or they simply lack important information that the reader needs to make critical decisions about what is being presented.

Pie charts

Pie charts are nice for showing how categorical data are broken down, but they can be misleading. Here's how to check a pie chart for quality:

>> Check to be sure the percentages add up to 100%, or close to it (any round-off error should be small).

>> Beware of slices labeled "Other" that are larger than the rest of the slices. This means the pie chart is too vague.

>> Watch for distortions with three-dimensional-looking pie charts, in which the slice closest to you looks larger than it really is because of the angle at which it's presented.

>> Look for a reported total number of individuals who make up the pie chart, so you can determine "how big" the pie is, so to speak. If the sample size is too small, the results are not going to be reliable.

Bar graphs

A bar graph breaks down categorical data by the number or percent in each group (see Chapter 3). When examining a bar graph:

>> Consider the units being represented by the height of the bars and what the results mean in terms of those units. For example, total number of crimes versus the crime rate (total number of crimes per capita).

>> Evaluate the appropriateness of the scale, or amount of space between units expressing the number in each group of the bar graph. Small scales (for example, going from 1 to 500 by 10s) make differences look bigger; large scales (going from 1 to 500 by 100s) make them look smaller.

Time charts

A time chart shows how some measurable quantity changes over time, for example, stock prices (see Chapter 3). Here are some issues to watch for with time charts:

>> Watch the scale on the vertical (quantity) axis as well as the horizontal (timeline) axis; results can be made to look more or less dramatic by simply changing the scale.

>> Take into account the units being portrayed by the chart and be sure they are equitable for comparison over time; for example, are dollars being adjusted for inflation?

>> Beware of people trying to explain why a trend is occurring without additional statistics to back themselves up. A time chart generally shows what is happening. *Why* it's happening is another story.

>> Watch for situations in which the time axis isn't marked with equally spaced jumps. This often happens when data are missing. For example, the time axis may have equal spacing between 1971, 1972, 1975, 1976, 1978, when it should actually show empty spaces for the years in which no data are available.

Histograms

Histograms graph numerical data in a bar-chart type of graph (seen in Chapter 3). Items to watch for regarding histograms:

>> Watch the scale used for the vertical (frequency/relative frequency) axis, especially for results that are exaggerated or played down through the use of inappropriate scales.

>> Check out the units on the vertical axis, whether they're reporting frequencies or relative frequencies, when examining the information.

>> Look at the scale used for the groupings of the numerical variable on the horizontal axis. If the groups are based on small intervals (for example, $0-2, 2-4$, and so on), the data may look overly volatile. If the groups are based on large intervals $(0-100, 100-200$, and so on), the data may give a smoother appearance than is realistic.

Biased Data

Bias in statistics is the result of a systematic error that either overestimates or underestimates the true value. Here are some of the most common sources of biased data:

>> Measurement instruments that are systematically off, such as a scale that always adds 5 pounds to your weight.

>> Participants that are influenced by the data-collection process. For example, the survey question, "Have you ever disagreed with the government?" will overestimate the percentage of people unhappy with the government.

>> A sample of individuals that doesn't represent the population of interest. For example, examining study habits by only visiting people in the campus library will create bias.

>> Researchers that aren't objective. Researchers have a vested interested in the outcome of their studies, and rightly so, but sometimes interest becomes influence over those results. For example, knowing who got what treatment in an experiment causes bias — double-blinding the study makes it more objective.

No Margin of Error

To evaluate a statistical result, you need a measure of its precision — that is, the margin of error (for example "plus or minus 3 percentage points"). When researchers or the media fail to report the margin of error, you're left to wonder about the accuracy of the results, or worse, you just assume that everything is fine, when in many cases it's not. Always check the margin of error. If it's not included, ask for it! (See Chapter 7 for all the details on margin of error.)

Nonrandom Samples

A random sample (as described in Chapter 12) is a subset of the population selected in such a way that each member of the population has an equal chance of being selected (like drawing names out of a hat). No systematic favoritism or exclusion is involved in a random sample. However, many studies aren't based on random samples of individuals; for example, TV polls asking viewers to "call us with your opinion"; an Internet survey you heard about from your friends; or a person with a clipboard at the mall asking for a minute of your time.

What's the effect of a nonrandom sample? Oh nothing, except it just blows the lid off of any credible conclusions the researcher ever wanted to make. Nonrandom samples are biased, and their data can't be used to represent any population beyond themselves. Check to make sure an important result is based on a random sample. If it isn't, run — and don't look back!

Missing Sample Sizes

Knowing how much data went into a study is critical. Sample size determines the precision (repeatability) of the results. A larger sample size means more precision, and a small sample size means less precision. Many studies (more than you would expect) are based on only a few subjects.

You might find that headlines and visual displays (such as graphs) are not exactly what they seem to be when the details reveal either a small sample size (reducing reliability in the results) or in some cases, no information at all about the sample size. For example, you've probably seen the chewing gum ad that says, "Four out of five dentists surveyed recommend [this gum] for their patients who chew gum." What if they really did ask only five dentists?

TECHNICAL STUFF

Always look for the sample size before making decisions about statistical information. Larger sample sizes have more precision than small sample sizes (assuming the data are of good quality). If the sample size is missing from the article, get a copy of the full report of the study or contact the researcher or author of the article.

Misinterpreted Correlations

Correlation is one of the most misunderstood and misused statistical terms used by researchers, the media, and the general public. (You can read all about this in Chapter 10.) Here are my three major correlation pet peeves:

>> **Correlation applies only to two *numerical* variables, such as height and weight.** So, if you hear someone say, "It appears that the voting pattern is correlated with gender,"

you know that's statistically incorrect. Voting pattern and gender may be associated, but they can't be correlated in the statistical sense.

>> **Correlation measures the strength and direction of a** *linear* **relationship.** If the correlation is weak, you can say there is no linear relationship; however, some other type of relationship might exist, for example, a curve (such as supply and demand curves in economics).

>> **Correlation doesn't imply cause and effect.** Suppose someone reports that the more people drink diet cola, the more weight they gain. If you're a diet cola drinker, don't panic just yet. This may be a freak of nature that someone stumbled onto. At most, it means more research needs to be done (for example, a well-designed experiment) to explore any possible connection.

Confounding Variables

Suppose a researcher claims that eating seaweed helps you live longer; you read interviews with the subjects and discover that they were all over 100, ate very healthy foods, slept an average of 8 hours a day, drank a lot of water, and exercised. Can you say the long life was caused by the seaweed? You can't tell, because so many other variables exist that could also promote long life (the diet, the sleeping, the water, the exercise); these are all confounding variables.

A common error in research studies is to fail to control for confounding variables, leaving the results open to scrutiny. The best way to head off confounding variables is to do a well-designed experiment in a controlled setting.

REMEMBER

Observational studies are great for surveys and polls, but not for showing cause-and-effect relationships, because they don't control for confounding variables. A well-designed experiment provides much stronger evidence. (See Chapter 13.)

Botched Numbers

Just because a statistic appears in the media doesn't mean it's correct. Errors appear all the time (by error or design), so look for them. Here are some tips for spotting botched numbers:

>> **Make sure everything adds up to what it's reported to.** With pie charts, be sure the percentages add up to 100% (or very close to it — there may be round-off error).

>> **Double-check even the most basic of calculations.** For example, a chart says 83% of Americans are in favor of an issue, but the report says 7 out of every 8 Americans are in favor of the issue. 7 divided by 8 is 87.5%.

>> **Look for the response rate of a survey — don't just be happy with the number of participants.** (The response rate is the number of people who responded divided by the total number of people surveyed times 100%.) If the response rate is much lower than 70%, the results could be biased, because you don't know what the nonrespondents would have said.

>> **Question the type of statistic used to determine if it's appropriate.** For example, the number of crimes went up, but so did population size. Researchers should have reported crime rate (crimes per capita) instead.

Statistics are based on formulas and calculations that don't know any better — the people plugging in the numbers should know better, though, but sometimes they either don't know better or they don't want you to catch on. You, as a consumer of information (also known as a certified skeptic), must be the one to take action. The best policy is to ask questions.

Selectively Reporting Results

Another bad move is when a researcher reports a "statistically significant" result but fails to mention that he found it among 50 different statistical tests he performed — the other 49 of which were *not* significant. This behavior is called *data fishing*, and that is not allowed in statistics. If he performs each test at a significance level of 0.05, that means he should expect to "find" a result that's not really there 5 percent of the time just by chance (see Chapter 8 for more on Type I errors). In 50 tests, he should expect at least one of these errors, and I'm betting that accounts for his one "statistically significant" result.

How do you protect yourself against misleading results due to data fishing? Find out more details about the study: How many tests were done, how many results weren't significant, and what was found to be significant? In other words, get the whole story if you can, so that you can put the significant results into perspective. You might also consider waiting to see whether others can verify and replicate the result.

The Almighty Anecdote

Ah, the anecdote — one of the strongest influences on public opinion and behavior ever created, and one of the least statistical. An anecdote is a story based on a single person's experience or situation. For example:

>> The waitress who won the lottery

>> The cat that learned how to ride a bicycle

>> The woman who lost 100 pounds on a potato diet

>> The celebrity who claims to use an over-the-counter hair color for which she is a spokesperson (yeah, right)

An anecdote is basically a data set with a sample size of one — they don't happen to most people. With an anecdote you have no information with which to compare the story, no statistics to analyze, no possible explanations or information to go on. You have just a single story. Don't let anecdotes have much influence over you. Rather, rely on scientific studies and statistical information based on large random samples of individuals who represent their target populations (not just a single situation).

Appendix
Tables for Reference

This appendix provides three commonly used tables for your reference: the Z-table, the t-table, and the Binomial table.

Because the first table won't fit on this page, I'd like to invite you to use this space to write down your innermost feelings about statistics.

TABLE A-1   The Z-Table

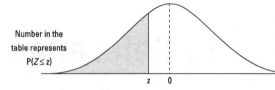

Number in the
table represents
$P(Z \leq z)$

z	0.00	0.01	0.02	0.03	0.04	0.05	0.06	0.07	0.08	0.09
-3.6	.0002	.0002	.0001	.0001	.0001	.0001	.0001	.0001	.0001	.0001
-3.5	.0002	.0002	.0002	.0002	.0002	.0002	.0002	.0002	.0002	.0002
-3.4	.0003	.0003	.0003	.0003	.0003	.0003	.0003	.0003	.0002	.0002
-3.3	.0005	.0005	.0005	.0004	.0004	.0004	.0004	.0004	.0003	.0003
-3.2	.0007	.0007	.0006	.0006	.0006	.0006	.0006	.0005	0005	.0005
-3.1	.0010	.0009	.0009	.0009	.0008	.0008	.0008	.0008	.0007	.0007
-3.0	.0013	.0013	.0013	.0012	.0012	.0011	.0011	.0011	.0010	.0010
-2.9	.0019	.0018	.0018	.0017	.0016	.0016	.0015	.0015	.0014	.0014
-2.8	.0026	.0025	.0024	.0023	.0023	.0022	.0021	.0021	.0020	.0019
-2.7	.0035	.0034	.0033	.0032	.0031	.0030	.0029	.0028	.0027	.0026
-2.6	.0047	.0045	.0044	.0043	.0041	.0040	.0039	.0038	.0037	.0036
-2.5	.0062	.0060	.0059	.0057	.0055	.0054	.0052	.0051	.0049	.0048
-2.4	.0082	.0080	.0078	.0075	.0073	.0071	.0069	.0068	.0066	.0064
-2.3	.0107	.0104	.0102	.0099	.0096	.0094	.0091	.0089	.0087	.0084
-2.2	.0139	.0136	.0132	.0129	.0125	.0122	.0119	.0116	.0113	.0110
-2.1	.0179	.0174	.0170	.0166	.0162	.0158	.0154	.0150	.0146	.0143
-2.0	.0228	.0222	.0217	.0212	.0207	.0202	.0197	.0192	.0188	.0183
-1.9	.0287	.0281	.0274	.0268	.0262	.0256	.0250	.0244	.0239	.0233
-1.8	.0359	.0351	.0344	.0336	.0329	.0322	.0314	.0307	.0301	.0294
-1.7	.0446	.0436	.0427	.0418	.0409	.0401	.0392	.0384	.0375	.0367
-1.6	.0548	.0537	.0526	.0516	.0505	.0495	.0485	.0475	.0465	.0455
-1.5	.0668	.0655	.0643	.0630	.0618	.0606	.0594	.0582	.0571	.0559
-1.4	.0808	.0793	.0778	.0764	.0749	.0735	.0721	.0708	.0694	.0681
-1.3	.0968	.0951	.0934	.0918	.0901	.0885	.0869	.0853	.0838	.0823
-1.2	.1151	.1131	.1112	.1093	.1075	.1056	.1038	.1020	.1003	.0985
-1.1	.1357	.1335	.1314	.1292	.1271	.1251	.1230	.1210	.1190	.1170
-1.0	.1587	.1562	.1539	.1515	.1492	.1469	.1446	.1423	.1401	.1379
-0.9	.1841	.1814	.1788	.1762	.1736	.1711	.1685	.1660	.1635	.1611
-0.8	.2119	.2090	.2061	.2033	.2005	.1977	.1949	.1922	.1894	.1867
-0.7	.2420	.2389	.2358	.2327	.2296	.2266	.2236	.2206	.2177	.2148
-0.6	.2743	.2709	.2676	.2643	.2611	.2578	.2546	.2514	.2483	.2451
-0.5	.3085	.3050	.3015	.2981	.2946	.2912	.2877	.2843	.2810	.2776
-0.4	.3446	.3409	.3372	.3336	.3300	.3264	.3228	.3192	.3156	.3121
-0.3	.3821	.3783	.3745	.3707	.3669	.3632	.3594	.3557	.3520	3483
-0.2	.4207	.4168	.4129	.4090	.4052	.4013	.3974	.3936	.3897	.3859
-0.1	.4602	.4562	.4522	.4483	.4443	.4404	.4364	.4325	.4286	.4247
-0.0	.5000	.4960	.4920	.4880	.4840	.4801	.4761	.4721	.4681	.4641

TABLE A-1 *(continued)*

Number in the
table represents
$P(Z \le z)$

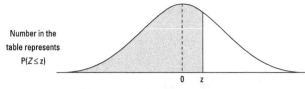

z	0.00	0.01	0.02	0.03	0.04	0.05	0.06	0.07	0.08	0.09
0.0	.5000	.5040	.5080	.5120	.5160	.5199	.5239	.5279	.5319	.5359
0.1	.5398	.5438	.5478	.5517	.5557	.5596	.5636	.5675	.5714	.5753
0.2	.5793	.5832	.5871	.5910	.5948	.5987	.6026	.6064	.6103	.6141
0.3	.6179	.6217	.6255	.6293	.6331	.6368	.6406	.6443	.6480	.6517
0.4	.6554	.6591	.6628	.6664	.6700	.6736	.6772	.6808	.6844	.6879
0.5	.6915	.6950	.6985	.7019	.7054	.7088	.7123	.7157	.7190	.7224
0.6	.7257	.7291	.7324	.7357	.7389	.7422	.7454	.7486	.7517	.7549
0.7	.7580	.7611	.7642	.7673	.7704	.7734	.7764	.7794	.7823	.7852
0.8	.7881	.7910	.7939	.7967	.7995	.8023	.8051	.8078	.8106	.8133
0.9	.8159	.8186	.8212	.8238	.8264	.8289	.8315	.8340	.8365	.8389
1.0	.8413	.8438	.8461	.8485	.8508	.8531	.8554	.8577	.8599	.8621
1.1	.8643	.8665	.8686	.8708	.8729	.8749	.8770	.8790	.8810	.8830
1.2	.8849	.8869	.8888	.8907	.8925	.8944	.8962	.8980	.8997	.9015
1.3	.9032	.9049	.9066	.9082	.9099	.9115	.9131	.9147	.9162	.9177
1.4	.9192	.9207	.9222	.9236	.9251	.9265	.9279	.9292	.9306	.9319
1.5	.9332	.9345	.9357	.9370	.9382	.9394	.9406	.9418	.9429	.9441
1.6	.9452	.9463	.9474	.9484	.9495	.9505	.9515	.9525	.9535	.9545
1.7	.9554	.9564	.9573	.9582	.9591	.9599	.9608	.9616	.9625	.9633
1.8	.9641	.9649	.9656	.9664	.9671	.9678	.9686	.9693	.9699	.9706
1.9	.9713	.9719	.9726	.9732	.9738	.9744	.9750	.9756	.9761	.9767
2.0	.9772	.9778	.9783	.9788	.9793	.9798	.9803	.9808	.9812	.9817
2.1	.9821	.9826	.9830	.9834	.9838	.9842	.9846	.9850	.9854	.9857
2.2	.9861	.9864	.9868	.9871	.9875	.9878	.9881	.9884	.9887	.9890
2.3	.9893	.9896	.9898	.9901	.9904	.9906	.9909	.9911	.9913	.9916
2.4	.9918	.9920	.9922	.9925	.9927	.9929	.9931	.9932	.9934	.9936
2.5	.9938	.9940	.9941	.9943	.9945	.9946	.9948	.9949	.9951	.9952
2.6	.9953	.9955	.9956	.9957	.9959	.9960	.9961	.9962	.9963	.9964
2.7	.9965	.9966	.9967	.9968	.9969	.9970	.9971	.9972	.9973	.9974
2.8	.9974	.9975	.9976	.9977	.9977	.9978	.9979	.9979	.9980	.9981
2.9	.9981	.9982	.9982	.9983	.9984	.9984	.9985	.9985	.9986	.9986
3.0	.9987	.9987	.9987	.9988	.9988	.9989	.9989	.9989	.9990	.9990
3.1	.9990	.9991	.9991	.9991	.9992	.9992	.9992	.9992	.9993	.9993
3.2	.9993	.9993	.9994	.9994	.9994	.9994	.9994	.9995	.9995	.9995
3.3	.9995	.9995	.9995	.9996	.9996	.9996	.9996	.9996	.9996	.9997
3.4	.9997	.9997	.9997	.9997	.9997	.9997	.9997	.9997	.9997	.9998
3.5	.9998	.9998	.9998	.9998	.9998	.9998	.9998	.9998	.9998	.9998
3.6	.9998	.9998	.9999	.9999	.9999	.9999	.9999	.9999	.9999	.9999

t-distribution showing area to the right

t (p, df)

df/p	0.40	0.25	0.10	0.05	0.025	0.01	0.005	0.0005
1	0.324920	1.000000	3.077684	6.313752	12.70620	31.82052	63.65674	636.6192
2	0.288675	0.816497	1.885618	2.919986	4.30265	6.96456	9.92484	31.5991
3	0.276671	0.764892	1.637744	2.353363	3.18245	4.54070	5.84091	12.9240
4	0270722	0.740697	1.533206	2.131847	2.77645	3.74695	4.60409	8.6103
5	0.267181	0.726687	1.475884	2.015048	2.57058	3.36493	4.03214	6.8688
6	0.264835	0.717558	1.439756	1.943180	2.44691	3.14267	3.70743	5.9588
7	0.263167	0.711142	1.414924	1.894579	2.36462	2.99795	3.49948	5.4079
8	0.261921	0.706387	1.396815	1.859548	2.30600	2.89646	3.35539	5.0413
9	0.260955	0.702722	1.383029	1.833113	2.26216	2.82144	3.24984	4.7809
10	0260185	0.699812	1.372184	1.812461	2.22814	2.76377	3.16927	4.5869
11	0259556	0.697445	1.363430	1.795885	2.20099	2.71808	3.10581	4.4370
12	0259033	0.695483	1.356217	1.782288	2.17881	2.68100	3.05454	43178
13	0.258591	0.693829	1.350171	1.770933	2.16037	2.65031	3.01228	4.2208
14	0.258213	0.692417	1.345030	1.761310	2.14479	2.62449	2.97684	4.1405
15	0.257885	0.691197	1.340606	1.753050	2.13145	2.60248	2.94671	4.0728
16	0257599	0.690132	1.336757	1.745884	2.11991	2.58349	2.92078	4.0150
17	0.257347	0.689195	1.333379	1.739607	2.10982	2.56693	2.89823	3.9651
18	0.257123	0.688364	1.330391	1.734064	2.10092	2.55238	2.87844	3.9216
19	0.256923	0.687621	1.327728	1.729133	2.09302	2.53948	2.86093	3.8834
20	0.256743	0.686954	1.325341	1.724718	2.08596	2.52798	2.84534	3.8495
21	0.256580	0.686352	1.323188	1.720743	2.07961	2.51765	2.83136	3.8193
22	0256432	0.685805	1.321237	1.717144	2.07387	2.50832	2.81876	3.7921
23	0256297	0.685306	1.319460	1.713872	2.06866	2.49987	2.80734	3.7676
24	0.256173	0.684850	1.317836	1.710882	2.06390	2.49216	2.79694	3.7454
25	0.256060	0.684430	1.316345	1.708141	2.05954	2.48511	2.78744	3.7251
26	0.255955	0.684043	1.314972	1.705618	2.05553	2.47863	2.77871	3.7066
27	0.255858	0.683685	1.313703	1.703288	2.05183	2.47266	2.77068	3.6896
28	0.255768	0.683353	1.312527	1.701131	2.04841	2.46714	2.76326	3.6739
29	0.255684	0.683044	1.311434	1.699127	2.04523	2.46202	2.75639	3.6594
30	0.255605	0.682756	1.310415	1.697261	2.04227	2.45726	2.75000	3.6460
∞	0.253347	0.674490	1.281552	1.644854	1.95996	2.32635	2.57583	3.2905

The Binomial Table

$$\text{Entry is } P(X = x) = \left(\begin{array}{c} n \\ x \end{array} \right) p^x (1-p)^{n-x}$$

n	x	0.1	0.2	0.25	0.3	0.4	0.5
2	0	0.810	0.640	0.563	0.490	0.360	0.250
	1	0.180	0.320	0.375	0.420	0.480	0.500
	2	0.010	0.040	0.063	0.090	0.160	0.250
3	0	0.729	0.512	0.422	0.343	0.216	0.125
	1	0.243	0.384	0.422	0.441	0.432	0.375
	2	0.027	0.096	0.141	0.189	0.288	0.375
	3	0.001	0.008	0.016	0.027	0.064	0.125
4	0	0.656	0.410	0.316	0.240	0.130	0.063
	1	0.292	0.410	0.422	0.412	0.346	0.250
	2	0.049	0.154	0.211	0.265	0.346	0.375
	3	0.004	0.026	0.047	0.076	0.154	0.250
	4	0.000	0.002	0.004	0.008	0.026	0.063
5	0	0.590	0.328	0.237	0.168	0.078	0.031
	1	0.328	0.410	0.396	0.360	0.259	0.156
	2	0.073	0.205	0.264	0.309	0.346	0.312
	3	0.008	0.051	0.088	0.132	0.230	0.312
	4	0.000	0.006	0.015	0.028	0.077	0.156
	5	0.000	0.000	0.001	0.002	0.010	0.031
6	0	0.531	0.262	0.178	0.118	0.047	0.016
	1	0.354	0.393	0.356	0.303	0.187	0.094
	2	0.098	0.246	0.297	0.324	0.311	0.234
	3	0.015	0.082	0.132	0.185	0.276	0.313
	4	0.001	0.015	0.033	0.060	0.138	0.234
	5	0.000	0.002	0.004	0.010	0.037	0.094
	6	0.000	0.000	0.000	0.001	0.004	0.016
7	0	0.478	0.210	0.133	0.082	0.028	0.008
	1	0.372	0.367	0.311	0.247	0.131	0.055
	2	0.124	0.275	0.311	0.318	0.261	0.164
	3	0.023	0.115	0.173	0.227	0.290	0.273
	4	0.003	0.029	0.058	0.097	0.194	0.273
	5	0.000	0.004	0.012	0.025	0.077	0.164
	6	0.000	0.000	0.001	0.004	0.017	0.055
	7	0.000	0.000	0.000	0.000	0.002	0.008

n	x	0.1	0.2	0.25	0.3	0.4	0.5
8	0	0.430	0.168	0.100	0.058	0.017	0.004
	1	0.383	0.336	0.267	0.198	0.090	0.031
	2	0.149	0.294	0.311	0.296	0.209	0.109
	3	0.033	0.147	0.208	0.254	0.279	0.219
	4	0.005	0.046	0.087	0.136	0.232	0.273
	5	0.000	0.009	0.023	0.047	0.124	0.219
	6	0.000	0.001	0.004	0.010	0.041	0.109
	7	0.000	0.000	0.000	0.001	0.008	0.031
	8	0.000	0.000	0.000	0.000	0.001	0.004
9	0	0.387	0.134	0.075	0.040	0.010	0.002
	1	0.387	0.302	0.225	0.156	0.060	0.018
	2	0.172	0.302	0.300	0.267	0.161	0.070
	3	0.045	0.176	0.234	0.267	0.251	0.164
	4	0.007	0.066	0.117	0.172	0.251	0.246
	5	0.001	0.017	0.039	0.074	0.167	0.246
	6	0.000	0.003	0.009	0.021	0.074	0.164
	7	0.000	0.000	0.001	0.004	0.021	0.070
	8	0.000	0.000	0.000	0.000	0.004	0.018
	9	0.000	0.000	0.000	0.000	0.000	0.002
10	0	0.349	0.107	0.056	0.028	0.006	0.001
	1	0.387	0.268	0.188	0.121	0.040	0.010
	2	0.194	0.302	0.282	0.233	0.121	0.044
	3	0.057	0.201	0.250	0.267	0.215	0.117
	4	0.011	0.088	0.146	0.200	0.251	0.205
	5	0.001	0.026	0.058	0.103	0.201	0.246
	6	0.000	0.006	0.016	0.037	0.111	0.205
	7	0.000	0.001	0.003	0.009	0.042	0.117
	8	0.000	0.000	0.000	0.001	0.011	0.044
	9	0.000	0.000	0.000	0.000	0.002	0.010
	10	0.000	0.000	0.000	0.000	0.000	0.001
12	0	0.282	0.069	0.032	0.014	0.002	0.000
	1	0.377	0.206	0.127	0.071	0.017	0.003
	2	0.230	0.283	0.232	0.168	0.064	0.016
	3	0.085	0.236	0.258	0.240	0.142	0.054
	4	0.021	0.133	0.194	0.231	0.213	0.121
	5	0.004	0.053	0.103	0.158	0.227	0.193
	6	0.000	0.016	0.040	0.079	0.177	0.226
	7	0.000	0.003	0.011	0.029	0.101	0.193
	8	0.000	0.001	0.002	0.008	0.042	0.121
	9	0.000	0.000	0.000	0.001	0.012	0.054
	10	0.000	0.000	0.000	0.000	0.002	0.016
	11	0.000	0.000	0.000	0.000	0.000	0.003
	12	0.000	0.000	0.000	0.000	0.000	0.000

TABLE A-3 *(continued)*

n	x	0.1	0.2	0.25	0.3	0.4	0.5
15	0	0.206	0.035	0.013	0.005	0.000	0.000
	1	0.343	0.132	0.067	0.031	0.005	0.000
	2	0.267	0.231	0.156	0.092	0.022	0.003
	3	0.129	0.250	0.225	0.170	0.063	0.014
	4	0.043	0.188	0.225	0.219	0.127	0.042
	5	0.010	0.103	0.165	0.206	0.186	0.092
	6	0.002	0.043	0.092	0.147	0.207	0.153
	7	0.000	0.014	0.039	0.081	0.177	0.196
	8	0.000	0.003	0.013	0.035	0.118	0.196
	9	0.000	0.001	0.003	0.012	0.061	0.153
	10	0.000	0.000	0.001	0.003	0.024	0.092
	11	0.000	0.000	0.000	0.001	0.007	0.042
	12	0.000	0.000	0.000	0.000	0.002	0.014
	13	0.000	0.000	0.000	0.000	0.000	0.003
	14	0.000	0.000	0.000	0.000	0.000	0.000
	15	0.000	0.000	0.000	0.000	0.000	0.000
20	0	0.122	0.012	0.003	0.001	0.000	0.000
	1	0.270	0.058	0.021	0.007	0.000	0.000
	2	0.285	0.137	0.067	0.028	0.003	0.000
	3	0.190	0.205	0.134	0.072	0.012	0.001
	4	0.090	0.218	0.190	0.130	0.035	0.005
	5	0.032	0.175	0.202	0.179	0.075	0.015
	6	0.009	0.109	0.169	0.192	0.124	0.037
	7	0.002	0.055	0.112	0.164	0.166	0.074
	8	0.000	0.022	0.061	0.114	0.180	0.120
	9	0.000	0.007	0.027	0.065	0.160	0.160
	10	0.000	0.002	0.010	0.031	0.117	0.176
	11	0.000	0.000	0.003	0.012	0.071	0.160
	12	0.000	0.000	0.001	0.004	0.035	0.120
	13	0.000	0.000	0.000	0.001	0.015	0.074
	14	0.000	0.000	0.000	0.000	0.005	0.037
	15	0.000	0.000	0.000	0.000	0.001	0.015
	16	0.000	0.000	0.000	0.000	0.000	0.005
	17	0.000	0.000	0.000	0.000	0.000	0.001
	18	0.000	0.000	0.000	0.000	0.000	0.000
	19	0.000	0.000	0.000	0.000	0.000	0.000
	20	0.000	0.000	0.000	0.000	0.000	0.000

Index

Symbols and Numerics

A

B

marijuana for chemotherapy side effects study, 149

matched-pairs design, 77, 99–102

maximum, in five-number summary, 21, 31

mean, 16–17, 34, 43. *See also* population mean (μ); sample mean

mean difference, testing, 99–102

median, 16–17, 20, 31, 34

minimum, in five-number summary, 21, 31

misleading question, 142–143

missed detection, 105–106

mistakes, common, 155–162

mothers in workforce examples, 25, 94–95

N

n (sample size). *See* sample size *(n)*

NBA salaries, 16–17

negative relationship, 114

New York Times, 149

n-factorial (*n*!), 39

nonrandom sample, 158–159

nonresponse to survey, minimizing, 140–142

normal distribution

about, 45–46

approximating to the binomial, 53–54

finding for given probability, 51–53

finding probabilities for, 48–50

standard (Z-distribution), 46–50, 108–109, 110

not-equal-to alternative hypothesis, 88–89, 91, 98–99, 111

null hypothesis (H$_o$), 88–90

numbers, botched, 161

numerical data, 9, 14–17

O

observational study, 5–6, 147–148, 160

ordinal data, 14

outcomes, defining in two-way table, 128

outlier, 16, 18, 32

overgeneralizing results, 11, 154

overstating results, 11, 154

P

p (probability). *See* probability *(p)*

paired test, 77, 99–102

Pearson correlation coefficient. *See* correlation *(r)*

percent, 14–15, 21

percentile, 18–20, 51–53, 109. *See also specific percentiles*

phone survey, 139

pie chart, 23–24, 156

pilot study, 74, 144

placebo, defined, 103

placebo effect, 150–151

poll, 74

population mean (μ)

comparing two, 78–80, 97–99

confidence interval for, 76–83, 111

testing one, 94–95

population proportion

confidence interval for, 77–78, 80–81

difference of two proportions, 80–81

testing one, 96–97

testing two, 102–104

population standard deviation (s), 60, 72–73

population variability, 75

positive relationship, 114

prediction, making, 124

probability *(p)*

conditional, 132–134

of failure, 38

finding for normal distribution, 48–50

finding for sample mean, 62–63

finding for sample proportion, 66–67

finding normal distribution for, 51–53

finding within two-way table, 130–134

joint, 130–131

marginal, 131–132

probability distribution, 35. *See also* binomial distribution

p-value, 92–94, 110–111

Q

Q1 (first quartile), 21, 31

Q3 (third quartile), 21, 31

qualitative data, 13–15

quantitative data, 14–17
question
  ACT math-help, 64–67
  wording of, 142–143

R

r (correlation). *See* correlation *(r)*
random selection, 7
random variable, 35–36, 43, 45
reading instruction methods example,
  99–102
regression line
  conditions, checking, 119
  cricket/temperature example, 123
  finding, 118–123
  formula, 120
  slope, finding, 120
  slope, interpreting, 121–122
  X- and *Y*-variables, 119
  y-intercept, finding, 121
  y-intercept, interpreting, 122–123
relationship
  cause-and-effect, 125–126, 160
  linear, 115, 160
  measuring using correlation,
    115–118
  negative, 114
  positive, 114
relative frequency, 8–9, 27, 30
reliability, 152
research hypothesis, 88–90
response rate, survey, 141, 161
response variable (*Y*-variable), 119
results
  overgeneralizing, 11, 154
  overstating, 11, 154
  selectively reporting, 161–162
rounding up sample size, 73–75

S

s (sample standard deviation), 17–18
saddle point, 46
sample
  defined, 138
  nonrandom, 158–159
  selecting, 7

target population, matching to,
  138–139
target population, randomly selecting from, 139–140
sample mean, 55, 56–57, 62–63.
  See also sampling distribution
sample proportion
  finding probabilities for, 66–67
  sampling distribution of, 63–66
sample size *(n)*
  bar graph, 25
  bias, compared to, 75
  confidence interval, effect on, 73–75
  experiment, 148–149
  formula, 73
  margin of error, effect on, 73–75
  missing, 159
  original, compared to final, 149
  pie chart, 156
  poll, 74
  rounding up, 73–75
  standard error, effect on, 58–59
  statistical meaning, 15
  survey, 139–140
sample standard deviation *(s)*, 17–18
sample statistic, 90–91
sampling distribution, 55–56, 60–67
sampling error, 83
sampling frame, 139
scale
  bar graph, 25, 156
  histogram, 30, 31, 157
  time chart, 26, 156
scatterplot, 114–115, 118
scientific survey, 144
score, standard, 90–91
selectively reporting results, 161–162
shape
  of data in histogram, 29
  of sampling distribution, 60–62
simple linear regression line.
  See regression line
skewed left data, 29, 32
skewed right data, 29, 32
slope *(m)*, 120, 121–122
standard deviation. *See also* variability
  population, 60, 72–73
  sample, 17–18
standard error, 57–62, 72–73

standard normal distribution
(Z-distribution), 46–50,
108–109, 110
standard score, 90–91
statistical analysis, 10, 153
statistics
 big-five, 121, 123
 descriptive, 8–9, 13–21
 as term, 2
statistics process overview
 conclusions, making, 10–11
 data, analyzing, 10
 data, collecting, 7–8
 data, describing, 8–9
 study, designing, 5–7
study design, 5–7
subject, experiment, 149–150
subjectivity, researcher, 158
survey
 anonymity, compared to confidenti-
 ality, 141–142
 benefits, 6
 bias, minimizing, 8
 checklist, 137–145
 CNN/Gallup, 142
 conclusions from, 144–145
 defined, 5–6
 disclaimer, 144
 following up, 140–141
 Internet, 140
 nonresponse, minimizing, 140–142
 phone, 139
 problems, potential, 6
 questions, wording of, 142–143
 response rate, 141, 161
 sample, matching to target population, 138–139
 sample, randomly selecting from
 target population, 139, 140
 sample, selecting, 7
 sample size, 139–140
 scientific, 144
 study design, 5–6
 target population, 138
 timing, 143
 training survey personnel, 143–144
 types, 5–6, 142
symmetric data, 32–33
symmetric distribution, 29

T

table. *See also* two-way table
 binomial, 40–42, 167–169
 t-, 109, 166
 Z-, 47–48, 52, 164–165
target population, 138–139
t-distribution
 about, 107–108
 confidence intervals with, 111
 critical values, finding, 110
 in hypothesis testing, 109–111
 p-value, finding, 110–111
 t-table, 109, 166
 Z-distribution, compared to,
 108–109, 110
test statistic, 90–94
third quartile (Q3), 21, 31
time chart, 26–27, 30, 156–157
traffic lights example, 39–42
treatment group, 150
t-table, 109, 166
t-value, 110
two-tailed hypothesis test, 110
two-way table
 about, 15
 conditional probabilities, calculating,
 132–134
 defined, 127
 independence, checking for,
 134–135
 joint probabilities, figuring,
 130–131
 marginal probabilities. calculating,
 131–132
 numbers, inserting, 129
 outcomes, defining, 128
 probabilities within, finding,
 130–134
 rows and columns, setting up,
 128–129
 totals, finding, 129–130
Type-1 error, 104–105
Type-2 error, 105–106

V

validity, 152
variability

About the Author

Deborah Rumsey is a Statistics Education Specialist and Auxiliary Professor at Ohio State University. Dr. Rumsey is a Fellow of the American Statistical Association and has won a Presidential Teaching Award from Kansas State University. She has served on the American Statistical Association's Statistics Education Executive Committee and the Advisory Committee on Teacher Enhancement, and is the editor of the Teaching Bits section of the *Journal of Statistics Education*. She is the author of the books *Statistics For Dummies, Statistics II For Dummies, Probability For Dummies, and Statistics Workbook For Dummies*. Her passions, besides teaching, include her family, fishing, birdwatching, getting "seat time" on her Kubota tractor, and cheering the Ohio State Buckeyes to another national championship.

Publisher's Acknowledgments

Project Editor: Corbin Collins

Senior Acquisitions Editor:
Lindsay Sandman Lefevere

Copy Editor: Corbin Collins

Assistant Editor:
Erin Calligan Mooney

Editorial Program Coordinator:
Joe Niesen

Technical Editors:
Jason J. Molitierno,
Jon-Lark Kim

Senior Editorial Manager:
Jennifer Ehrlich

**Editorial Supervisor and Reprint
Editor:** Carmen Krikorian

Editorial Assistants: Rachelle Amick,
Jennette ElNaggar

Senior Editorial Assistant:
David Lutton

Production Editor:
Magesh Elangovan

Cover Image: © enot-poloskun/
Getty Images

Take dummies with you everywhere you go!

Whether you are excited about e-books, want more from the web, must have your mobile apps, or are swept up in social media, dummies makes everything easier.

PERSONAL ENRICHMENT

Staying Sharp
9781119187790
USA $26.00
CAN $31.99
UK £19.99

Facebook
9781119179030
USA $21.99
CAN $25.99
UK £16.99

Guitar
9781119293354
USA $24.99
CAN $29.99
UK £17.99

Investing
9781119293347
USA $22.99
CAN $27.99
UK £16.99

Beekeeping
9781119310068
USA $22.99
CAN $27.99
UK £16.99

Digital Photography
9781119235606
USA $24.99
CAN $29.99
UK £17.99

Meditation
9781119251163
USA $24.99
CAN $29.99
UK £17.99

Pregnancy
9781119235491
USA $26.99
CAN $31.99
UK £19.99

Samsung Galaxy S7
9781119279952
USA $24.99
CAN $29.99
UK £17.99

iPhone
9781119283133
USA $24.99
CAN $29.99
UK £17.99

Crocheting
9781119287117
USA $24.99
CAN $29.99
UK £16.99

Nutrition
9781119130246
USA $22.99
CAN $27.99
UK £16.99

PROFESSIONAL DEVELOPMENT

Windows 10
9781119311041
USA $24.99
CAN $29.99
UK £17.99

AutoCAD
9781119255796
USA $39.99
CAN $47.99
UK £27.99

Excel 2016
9781119293439
USA $26.99
CAN $31.99
UK £19.99

QuickBooks 2017
9781119281467
USA $26.99
CAN $31.99
UK £19.99

macOS Sierra
9781119280651
USA $29.99
CAN $35.99
UK £21.99

LinkedIn
9781119251132
USA $24.99
CAN $29.99
UK £17.99

Windows 1
9781119310560
USA $34.00
CAN $41.99
UK £24.99

SharePoint 2016
9781119181705
USA $29.99
CAN $35.99
UK £21.99

Fundamental Analysis
9781119263593
USA $26.99
CAN $31.99
UK £19.99

Networking
9781119257769
USA $29.99
CAN $35.99
UK £21.99

Office 2016
9781119293477
USA $26.99
CAN $31.99
UK £19.99

Office 365
9781119265313
USA $24.99
CAN $29.99
UK £17.99

Salesforce.com
9781119239314
USA $29.99
CAN $35.99
UK £21.99

Coding
9781192933
USA $29.99
CAN $35.99
UK £21.99

CPSIA information can be obtained
at www.ICGtesting.com
Printed in the USA
LVHW080849160920
665975LV00025B/1538